AF361867

Conservation Tools, Protocols and Treatments on Painted Surfaces, Metal Leaves and Finishes in Cultural Heritage

Conservation Tools, Protocols and Treatments on Painted Surfaces, Metal Leaves and Finishes in Cultural Heritage

Editor

Marco Nervo

MDPI • Basel • Beijing • Wuhan • Barcelona • Belgrade • Manchester • Tokyo • Cluj • Tianjin

Editor
Marco Nervo
Fondazione Centro per la
Conservazione ed il Restauro dei
Beni Culturali
"La Venaria Reale"
Italy

Editorial Office
MDPI
St. Alban-Anlage 66
4052 Basel, Switzerland

This is a reprint of articles from the Special Issue published online in the open access journal *Coatings* (ISSN 2079-6412) (available at: https://www.mdpi.com/journal/coatings/special_issues/painted_surfaces).

For citation purposes, cite each article independently as indicated on the article page online and as indicated below:

LastName, A.A.; LastName, B.B.; LastName, C.C. Article Title. *Journal Name* **Year**, *Volume Number*, Page Range.

ISBN 978-3-0365-3254-7 (Hbk)
ISBN 978-3-0365-3255-4 (PDF)

Contents

About the Editor

Marco Nervo (Ph.D.) graduates in physics and takes his Ph.D. at the University of Torino, he conducted his studies in the Compact Muon Solenoid experiment at CERN in Geneva, developing an experimental thesis on the construction of particle detectors. Since 2007 he worked in the Foundation Centro per la Conservazione ed il Restauro dei Beni Culturali "La Venaria Reale" and since 2011 he is Head of Scientific Laboratories. He deals mainly with non-invasive diagnostic techniques, in particular XRF, digital radiographs, computerized tomography, thermo-hygrometric monitoring and software development for data analysis.

Starting from 2020 he leaves the role of Head of Scientific Laboratories to take on the role of Head of Purchasing Department, Technical Department, IT, Quality Management System.

Editorial

Conservation Tools, Protocols, and Treatments on Painted Surfaces, Metal Leaves, and Finishes in Cultural Heritage

Marco Nervo

Fondazione Centro per la Conservazione ed il Restauro dei Beni Culturali "La Venaria Reale", 10078 Venaria Reale, TO, Italy; marco.nervo@centrorestaurovenaria.it

Citation: Nervo, M. Conservation Tools, Protocols, and Treatments on Painted Surfaces, Metal Leaves, and Finishes in Cultural Heritage. *Coatings* **2022**, *12*, 164. https://doi.org/10.3390/coatings12020164

Received: 21 January 2022
Accepted: 26 January 2022
Published: 27 January 2022

Publisher's Note: MDPI stays neutral with regard to jurisdictional claims in published maps and institutional affiliations.

The conservation of painted surfaces, metal leaves, and finishes requires a deep knowledge of both the materials themselves and the supports, in addition to the interaction phenomena occurring among them. Superficial treatments, operations, and materials adopted during the conservation intervention can modify the complex system of existing interactions. Therefore, it is fundamental to predict possible induced changes in the chemical–physical properties of the systems.

Diagnostic techniques can make a huge contribution to conservation treatments of works of art. Rather than focusing on a single analysis or pure analytical data related to materials characterisation, we think it is interesting to focus on decay processes and reciprocal interactions involving the analysed materials. From a conservative point of view, the study of the interactions of the materials with each other and with the environment is, in fact, often more important than the characterisation of the materials themselves.

Defining a study methodology aimed at providing the correct technical–scientific support to conservation treatments therefore becomes essential. Another fundamental theme is also linked to this: how can I evaluate the effectiveness of a conservation treatment and the recognisability of the treatments themselves? The definition of protocols for the evaluation and validation of these treatments also in this case requires a further step with respect to the mere characterisation of the materials.

Therefore, there is then the complex theme of new materials and their use in treatments on works of art. I know the chemical–physical characteristics of these materials, but do I also know their aging mechanisms over time? What will be the interaction mechanisms between ancient and modern materials when the latter have aged?

This Special Issue aims at contributing to the definition of the state of the art in the approach to conservation problems of painted surfaces, metal leaves, and finishes.

Some of the papers in this Special Issue show how, through a complex diagnostic campaign, it is possible to provide useful elements for choosing the best and most suitable surface treatment methods. Others, always starting from analytical investigations, highlight the possible evolutions that the materials may have over time and the decay processes that they may undergo. The evaluation of the effectiveness of the treatments is another of the topics addressed, in particular in regard to the use of lasers in cultural heritage.

Funding: This research received no external funding.

Institutional Review Board Statement: Not applicable.

Informed Consent Statement: Not applicable.

Conflicts of Interest: The authors declare no conflict of interest.

 *coatings*

Article

An Ancient Egyptian Multilayered Polychrome Wooden Sculpture Belonging to the Museo Egizio of Torino: Characterization of Painting Materials and Design of Cleaning Processes by Means of Highly Retentive Hydrogels

Nicole Manfredda [1], Paola Buscaglia [1,*], Paolo Gallo [2], Matilde Borla [3], Sara Aicardi [4], Giovanna Poggi [5], Piero Baglioni [5], Marco Nervo [1,9], Dominique Scalarone [7], Alessandro Borghi [8], Alessandro Re [6,9] Laura Guidorzi [6,9] and Alessandro Lo Giudice [6,9]

[1] Centro Conservazione e Restauro la Venaria Reale, 10078 Venaria Reale, Italy; nicole.manfredda@edu.unito.it (N.M.); marco.nervo@centrorestaurovenaria.it (M.N.)
[2] Dipartimento di Studi Storici, Università di Torino, 10124 Torino, Italy; p.gallo@unito.it
[3] Soprintendenza Archeologia, Belle Arti e Paesaggio per la città Metropolitana di Torino, 10124 Torino, Italy; matilde.borla@beniculturali.it
[4] Museo Egizio di Torino, 10124 Torino, Italy; sara.aicardi@museoegizio.it
[5] Dipartimento di Chimica and CSGI, Università di Firenze, 50019 Sesto Fiorentino, Italy; poggi@csgi.unifi.it (G.P.); baglioni@csgi.unifi.it (P.B.)
[6] Dipartimento di Fisica, Università di Torino, 10124 Torino, Italy; alessandro.re@unito.it (A.R.); laura.guidorzi@unito.it (L.G.); alessandro.logiudice@unito.it (A.L.G.)
[7] Dipartimento di Chimica, Università di Torino, 10124 Torino, Italy; dominique.scalarone@unito.it
[8] Dipartimento di Scienze della Terra, Università di Torino, 10124 Torino, Italy; alessandro.borghi@unito.it
[9] Istituto Nazionale di Fisica Nucleare, Sezione di Torino, 10124 Torino, Italy
[*] Correspondence: paola.buscaglia@centrorestaurovenaria.it; Tel.: +39-011-499-3060

Citation: Manfredda, N.; Buscaglia, P.; Gallo, P.; Borla, M.; Aicardi, S.; Poggi, G.; Baglioni, P.; Nervo, M.; Scalarone, D.; Borghi, A.; et al. An Ancient Egyptian Multilayered Polychrome Wooden Sculpture Belonging to the Museo Egizio of Torino: Characterization of Painting Materials and Design of Cleaning Processes by Means of Highly Retentive Hydrogels. *Coatings* 2021, 11, 1335. https://doi.org/10.3390/coatings11111335

Academic Editor: Robert J. K. Wood

Received: 11 September 2021
Accepted: 15 October 2021
Published: 30 October 2021

Publisher's Note: MDPI stays neutral with regard to jurisdictional claims in published maps and institutional affiliations.

Abstract: This contribution focuses on the conservation of an Egyptian wooden sculpture (Inventory Number Cat. 745) belonging to the Museo Egizio of Torino in northwest Italy. A preliminary and interdisciplinary study of constituent painting materials and their layering is here provided. It was conducted by means of a multi-technique approach starting from non-invasive multispectral analysis on the whole object, and subsequently, on selected micro-samples. In particular, visible fluorescence induced by ultraviolet radiation (UVF), infrared reflectography (IRR) and visible–induced infrared luminescence were used on the whole object. The micro-samples were analysed by means of an optical microscope with visible and UV light sources, a scanning electron microscope (SEM) with an energy-dispersive X-ray spectrometer (EDX), Fourier transform infrared (FT-IR) spectrometer, pyrolysis-gas chromatography/mass spectrometer (Py-GC/MS) and micro-particle induced X-ray emission (PIXE). The characterization of the painting materials allowed the detection of Egyptian blue and Egyptian green, and also confirmed the pertinence of the top brown layer to the original materials, which is a key point to design a suitable surface treatment. In fact, due to the water sensitiveness of the original materials, only few options were available to perform cleaning operations on this artwork. To setup the cleaning procedure, we performed several preliminary tests on mockups using dry cleaning materials, commonly used to treat reactive surfaces, and innovative highly water retentive hydrogels, which can potentially limit the mechanical action on the original surface while proving excellent cleaning results. Overall, this study has proved fundamental to increase our knowledge on ancient Egyptian artistic techniques and contribute to hypothesize the possible provenance of the artefact. It also demonstrated that polyvinyl alcohol-based retentive gels allow for the safe and efficient cleaning of extremely water sensitive painted surfaces, as those typical of ancient Egyptian artefacts.

Keywords: cultural heritage; conservation; wooden sculpture; ancient Egyptian; ancient Egyptian painting materials; cleaning treatment; water based systems; poly(vinyl alcohol) hydrogels; archaeometry

1. Introduction

Designing a correct conservation treatment process requires an interdisciplinary work that involves professionals from various fields, in order to combine and incorporate several approaches and data to reach an overview, as complete as possible, of the artefact. In particular, for archaeological objects, besides notes on the specific excavation activities, few documented information are available on the artistic techniques and conservation history. In this context, archaeometric studies do strongly contribute to the study of archaeological finds.

This paper focuses on the study of an Egyptian wood sculpture dating back to New Kingdom (1550–1069 B.C.E.), which belongs to the Museo Egizio of Torino (Inventory Number Cat. 745). In particular, the characterization of painting materials and the set-up of an adequate cleaning treatment of the surface of the sculpture are here provided.

The Cat. 745 statuette represents Hapy, god of the Nile flood, or a more generic fecundity figure, and seems to be the only known case, until now, of a wooden sculpture depicting this subject (Figure 1). The figure is standing, placed on a rectangular base, with the left feet ahead. It is missing of his forearms that probably held an offering table, in consideration of both the morphology of joints and of the common iconography of these figures. It is represented with a wig and dressed only with a belt fixed on the hips, characterized by three long stripes up to the knees. This type of figures often has an insignia on their head representing their name but, in this case, this feature is missing, making impossible to ascertain the specific representation [1].

Figure 1. Inventory Number Cat. 745 before the cleaning treatment. Measurements (h × w × d) 62 cm × 16 cm × 25.5 cm. (**a**): front side, diffuse light. (**b**): back side, diffuse light.

The statuette was studied from 2005 to 2007 within the framework of three Bachelor theses [2–4] and, later (2017–2019), in the framework of a Master dissertation [5], which focused on conservation treatments. Taking into account the unusual multi-layered polychromies, scientific insights have been provided, in order to characterize both pigments and binders for a better comprehension of the artistic technique. Moreover, it appeared necessary to detect the presence of non-original materials, in the view of setting up a cleaning treatment, for a proper reading of the decoration of the statuette.

A brief description of the case study and of its layering of materials seems necessary to understand the aim of the research process.

In terms of painted decoration, the artefact presents, by a visual inspection, multi-layered polychromies, alternating ground layers and coloured layers. Specifically, in the body and wig parts of the sculpture five layers of materials are present (Figure 2), two of which refer to preparation layers (in Figure 2, starting from the interface with the wooden material, layer 1 and layer 3), while the other three correspond to paint layers.

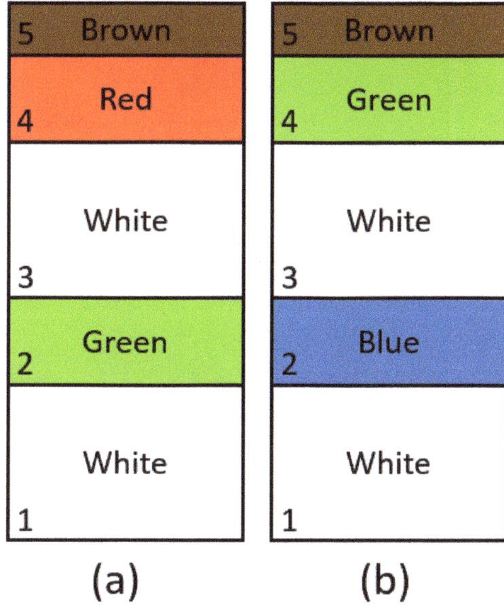

Figure 2. Scheme of the layer sequence in painting materials in the wig (**a**) and in the body (**b**) of the sculpture. The thickness of the layers is not in scale.

Having no information about the history of the object before its arrival in Torino, we could only presume that it was acquired between 1824 and 1882, the year in which it is described in a catalogue [6], with the inventory number (745) and its original location. Few information is also available about its conservation history. Archive black & white photos dating back to the 1970s were retrieved; they showed a thick layer of dust, visible in particular on the upper face of the base. In addition to this, the only documented report on a securing treatment carried on the artefact by Luigi Vigna, at that time director of the conservation laboratory of the Museo Egizio, was found [2–4]. In this context, the removal of the dust and the consolidation of the preparation and paint were carried out, with the aim of allowing a correct reading and a safe handling of the object. Further maintenance interventions, which are difficult to place chronologically have been indeed carried out: for instance a modern nail inserted to reinforce the assembly of the arm was clearly detected by X-ray radiography (not shown here).

Considering the complex layering of materials, it was necessary to understand if the cleaning operations had to be limited to the surface dirt, or if the removal of one or more paint layers might be considered, if these were demonstrated to be non-original. In particular, the characterization of the brown top layer seemed necessary: even if not visible near to the gaps, and therefore, apparently not applied after some deterioration, without more information on its nature, it was not possible to exclude in advance its pertinence to a non-original patina. Moreover, it is worth noting that the superficial dirt was placed on top of a rough, porous and extremely fragile surface, which could have made the cleaning operation particularly challenging (Figure 3).

Figure 3. Detail of the back of the case study. It can be seen that different paint layers are covered with dust.

Archaeological painting materials as the Egyptian ones, are often strongly water-sensitive, and cannot be cleaned with traditional gels such as agarose-based ones. However, in consideration of the lack of superficial cohesion of the materials, any mechanical action, as that associated with dry cleaning methods, might have been too stressful for the surface. Therefore, it was decided to compare traditional dry-cleaning methods, widely tested and commonly used on archaeological polychromies [7], with innovative highly retentive water-based hydrogels, recently introduced in conservation practice [8–10]. In particular, among the available formulations, it was decided to test twin-chain polymer hydrogels based on poly(vinyl alcohol), which have been developed for the cleaning of water-sensitive modern and contemporary artworks [11–14]. These systems, which, to the best of our knowledge, have not been previously tested on ancient Egyptian artefacts, combine good adhesion to rough and textured paint layers, and controlled wetting of surfaces, granting safe removal of soil. The combination of dry-cleaning materials with highly retentive gels was expected to grant a delicate and localized action at the interface with the original paint layer.

2. Materials and Methods

2.1. Painting Materials Analysis

To identify the used pigments and to study the complex structure of painting materials, an approach based on non-invasive and micro-invasive techniques approach in a two-step sequence was applied as provided in standard protocols for conservation purposes [15], starting from non-invasive multispectral analysis on the whole object, and subsequently, on selected samples.

For the characterization of painting materials three typologies of non-invasive analysis techniques were employed: visible fluorescence induced by ultraviolet radiation (UVF), infrared reflectography (IRR) and visible-induced infrared luminescence (VIL). UVF is useful for the identification of different materials on the surface which are not easily discerned using visible light, including previous conservation treatments. IRR uses the near infrared to investigate under the paint layers and to highlight the potential presence of preparatory drawings or other pictorial materials. Moreover, it can be combined with visible light photographs to produce false colour (IRFC) images, which can be used to tentatively identify pigments. Finally, the VIL technique is mainly used to identify the presence of Egyptian blue. In all the cases the Adobe Photoshop (Adobe, San Jose, CA, USA) software was used for the post-production of images.

UVF images were captured by means of a Xnite Nikon D810 camera (Nikon Corporation, Tokyo, Japan) coupled with a PECA 916 digital filter (Peca Products Inc., Beloit, WI, USA). The illumination was obtained by means of two UV Labino® spot lamps UV light MPXL and UV FLOODLIGHT (Labino, Stockholm, Sweden) with emission peak at 365 nm. A white standard Spectralon® (Labsphere Inc., NH, USA, nominal reflectante: 99% in the field of imaging) was used for the white balance.

To carry out IRR images, the sculpture was illuminated by means of two 800W Varibeam halogen lamps (Ianiro, Eagle Beaming International Co. Ltd., New Taipei City, Taiwan) and photographed by means of the Xnite Nikon D810 camera equipped with an infrared filter (R72, Hoya, HOYA CORPORATION, Tokyo, Japan) for detecting the wavelengths in the range 780–950 nm. In this case a ColorChecker® Classic 24 colours (X-Rite, MI 49512, Grand Rapids, MI, USA) placed in the field of imaging was considered for the chromatic balance.

VIL images were acquired using the same Xnite Nikon D810 (digital camera but equipped with a Peca 910 filter (Peca Products Inc., Beloit, WI, USA, 750–950 nm). The illumination was obtained by means of a LED light with 400–700 nm emission and a Peca 916 filter placed on it. As references, both a ColorChecker® Classic 24 colours and a pad of Egyptian blue (Kremer Pigment n° 10060) were used.

To better understand the composition of painting materials and to contribute in confirming the relevance of the surface layer to the original materials, micro-invasive analysis were performed on micro-samples. Samplings were made in significant areas of the sculpture. In particular, for the stratigraphic study small pieces were taken from the body (layers sequence is shown in Figure 2a) and from the wig (layers sequence is shown in Figure 2b). Moreover, a small amount of powder was extracted from the white belt and from the brown paint covering all the surface of the sculpture. To minimize the invasiveness, no samplings were made on yellow decorations, on the black lines on the wig and on the basement because they were considered not significant for conservation aims.

The samples from body and wig were prepared in a polished section and were observed by means of a BX51 minrepetrographic microscope (Olympus Corporation, Shinjuku, Tokyo, Japan) in visible and UV light, interfaced to a PC by means of a digital camera. The acquisition and processing of images is carried out using the proprietary software analySIS Five.

SEM-EDX measurements were performed in order to determine the chemical composition of the main minerals. A JSM-IT300LV scanning electron microscope (JEOL, Tokyo, Japan) equipped with an energy-dispersive X-ray spectrometer (EDX), with a SDD (Oxford Instruments, Oxford, UK), hosted at the Earth Science Department of the University of

Torino, was used for the determination of major elements. The measurements were conducted in high vacuum conditions after covering of the sample surface with a conductive layer. Spot analyses were acquired under the following conditions: accelerating voltage 15 kV, counting time 50 s, process time 5 μs and working distance 10 mm. The EDX-acquired spectra were corrected and calibrated both in energy and in intensity thanks to measurements performed on cobalt standard introduced in the vacuum chamber with the samples. The Microanalysis Suite Oxford INCA Energy 200 (Oxford Instruments, Abingdon, Oxfordshire, UK) that enables spectra visualization and elements recognition, was employed. A ZAF data reduction program was used for spectra quantification. The resulting full quantitative analysis was obtained from the spectra, using natural oxides and silicates from Astimex Scientific Limited® (Astimex, Kista, Sweden) as standards. All the analyses were recalculated using the MINSORT computer software [16].

Preliminary ion beam analyses (IBA) measurements, in particular micro-particle induced X-ray emission (PIXE) tests were carried out in order to determine the presence of minor and trace elements in correlation with each pigment layer. This information could be useful to characterize the material and to find markers helpful in to identifying different productions in time. IBA measurements were carried out at the micro-beam line of the Legnaro National Laboratory (INFN-LNL) located in Padova (Italy), using a 2 MeV proton beam. Because samples are placed in a vacuum chamber for the measurements, they were prepared following a procedure reported elsewhere [17], which is similar to the one used for SEM anaysis.

Fourier transform infrared (FT-IR) spectroscopy was carried out to characterize the white pigment of the belt and brown surficial layer. The measurements were conducted on selective micro-samples with a Vertex 70 FT-IR spectrophotometer (Bruker, Billerica, MA, USA) coupled with a Bruker Hyperion infrared microscope working in transmission mode with the aid of a diamond cell.

On a sample from the brown layer, pyrolysis-gas chromatography/mass spectrometry (Py-GC/MS) was performed with an EGA/PY-3030D pyrolyzer (Frontier Lab, Koriyama, Fukushima, Japan) interfaced with a 6890N Network GC System (Agilent Technologies, Wilmington, DE, USA) with HP-5 cross-linked 5% Ph Me silicone capillary column (30 m × 0.25 mm × 0.25 μm) and a 5973 Network Mass Selective Detector (Agilent Technologies). The sample was derivatized with the thermally assisted hydrolysis and methylation (THM) method using tetramethylammonium hydroxide (TMAH) in aqueous solution at a concentration of 25% by weight (Sigma-Aldrich, Milan, Italy). Pyrolysis was carried out at 650 °C for 12 s. The interface temperature of the pyrolyzer and of the injector of the gas chromatograph was 300 °C. The following temperature program was used for the gas chromatographic separation: isotherm of 2 min at 50 °C, ramp of 10 °C/min up to 300 °C, isotherm at 300 °C for 5 min. The carrier gas was helium (1.0 mL/min) and split ratio was 1/20 of the total flow. Mass spectra were recorded under electron impact at 70 eV, scan range 40–650 m/z. The interface was kept at 280 °C, ion source at 230 °C and quadrupole mass analyzer at 150 °C. All instruments were controlled by Enhanced Chem Station (ver. 9.00.00.38) software. The mass spectra assignment was done with the Wiley 138 and NIST2008 libraries and by comparison with literature data.

2.2. Mockups for Cleaning Test

2.2.1. Mockups

On the basis of the results on pigment characterization (described in Section 3.1), seven different layers of colours have been prepared on glass slides, in five replicas (series). A picture of the samples as prepared is shown in Figure 4a. Due to the presence of losses and superficial abrasions, experimental preliminary activity to define the better cleaning treatment was carried out on mockups featuring also the paints underneath the brown top layer (samples A–D in Figure 4a). Samples E, F and G, which mimic the polychromies underneath the brown top layer, have been used to test the effect of different cleaning procedures on surfaces with variable roughness underneath the brown layer. Commercial products were used to prepare the mockups, even if we are aware of the unlikeliness of a real replicability of the original materials, with particular reference to proper production processes and aging. Concerning the reproduction of the brown layer, due to its complexity and in consideration of the main aims of the experimental activity, a mixture of earth pigments chromatically comparable to the original brown was considered adequate. For all the samples, extra quality gum Arabic (Bresciani Srl, Milan, Italy) at 10% in water and micronized calcium carbonate (CTS Europe, Milan, Italy) were used as a binder for paint layers and for preparation layer, respectively. Mockups' composition is summarized in Table 1.

Figure 4. Mockups (series 5) used for cleaning trials, before (**a**) and after the soiling procedure (**b**). In the soiled samples, the reference area is covered with a white paper tissue.

Table 1. Mockups' composition. Preparation layer is calcium carbonate with gum Arabic as binder (10% solution in water), Brown is a mixture of 50% Burnt umber (Z.C0798), 25% Red ochre (Z.C0008) and 25% Green earth (Z.C0320).

Model Sample	Pigments of Paint Layer [1]	Layering of Materials [1]
A	None	Preparation layer
B	Egyptian blue (KP.10060)	Preparation layer-KP.10060
C	Egyptian green (KP.10064)	Preparation layer-KP.10064
D	Yellow ochre (KP.40301) + Red ochre (Z.C0008)	Preparation layer-KP.40301-Z.C0008
E	Yellow ochre (KP.40301) + Black (S.7060030) + Brown	Preparation layer-KP.40301-S.7060030 (lines)-Brown
F	Red ochre (Z.C0008) + Black (S.7060030) + Brown	Preparation layer-Z.C0008-S.7060030 (stripes)-Brown
G	Egyptian green (KP.10064) + Black (S.7060030) + Brown	Preparation layer-KP.10064-S.7060030 (lines)-Brown

[1] Pigment manufacturers: KP = Kremer Pigment; Z = Zecchi; S = Sinopia.

A first step of artificial ageing has been foreseen to simulate the oxidation and degradation of the binders The samples have been aged for 500 h in a simulating solar irradiation solar box Heraeus Suntest CPS (Heraeus Holding GmbH, Hanau, Germany) equipped with a filtered (coated quartz glass simulating a 3 mm window glass, cutting l < 300 nm) xenon lamp and with an average irradiation of 750 W/m^2 and an internal temperature of about 50 °C.

After the first step of ageing, we simulated the fatty superficial dirt detected on the artefact (Figure 4b), by applying a thin layer of INCI hands cream (Yves Rocher®, now Rocher Group®, Rennes, France) to obtain a greashy surface capable of incorporating dust.

Dust was recovered from the filter of a museum vacuum cleaner commonly used in maintenance operations by Centro Conservazione e Restauro La Venaria Reale professionals. On each sample, a small area was covered during the application to be used as a reference. After applying a thick coat of dust, samples were placed in a humidity non-watertight craft chamber, realized with a wooden structure covered with Melinex® foils. Water at 25 °C was nebulized until saturation, and samples were left there for 4 h, i.e., the time needed to return to the outside environmental conditions. This procedure allowed dust to better adhere to the surfaces. Afterwards, dust excess not firmly attached to the surface was removed shaking the mockups upside down and using a compressed air jet by means of an airbrush.

2.2.2. Selected Cleaning Materials

Latex free high-density polyurethane (PU) sponges (Deffner&Johann®, Röthlein, Germany) have been compared with innovative highly retentive hydrogels. The use of this specific PU sponges is common in the professional practice in case of archaeological materials, with procedures that follow the results of a European research project specifically dedicated to dry-cleaning [8].

Among the available innovative highly retentive hydrogels recently introduced in conservation practice, twin-chain polymer hydrogels based on poly(vinyl alcohol), developed within the H2020 European project NANORESTART (grant agreement 646063), were selected. One of the most interesting features of these systems is their capability of adapting to the three-dimensional objects and irregular surfaces, such as the painted areas of the Egyptian sculpture. Moreover, the hydrophilic poly(vinyl alcohol)-based structural network makes the gel capable of holding large amounts of aqueous liquid, while the highly retentive properties limit the liquid's penetration so that cleaning occurs only at the interface, without affecting the surrounding area or leaving residues. Within the European project, formulations were tailored to adapt to the specific requirements of several case studies, but also served as prototypes for a series of multipurpose gels, which were formulated to target typical cleaning cases. Gels are named Nanorestore Gels® Peggy 5 and Nanorestore Gels® Peggy 6 (CSGI, Florence, Italy), being the first more retentive and rigid than the second. Both formulations are available in different shapes, including thin foils (sheets) or in parallelepiped shape (gum).

After some preliminary test, Nanorestore Gel® Peggy 6 (sheets, PG6) and Nanorestore Gel® Peggy Gum 5 (PG5 Gum) were selected. The first are more flexible and easily adapt to the artefact. The second ones were selected due to their shape that provides an easier handling and allow for a gentle and punctual mechanical action, possibly increasing the efficacy of the cleaning.

With the aim of defining the best cleaning procedure, different combinations of materials and application lengths were tested, as summarized in Table 2.

Table 2. Cleaning tests carried out on mockups. For each sample a non-soiled area has been kept as reference.

Materials Tested on Each Sample	Objective	Test Name	Test Description
PG6	Tuning the length of gel sheet's application	2a	180 s
		2b	150 s
		2c	120 s + 60 s
		3a	90 s + 90 s
		3b	60 s + 60 s
		3c	30 s
PG6 PG5 Gum	Comparing the effect of gel gums applied on wet and dry surfaces	4a	PG6 (90 s + 90 s) + PG5 Gum on a still wet surface
		4b	PG6 (90 s + 90 s) + PG5 Gum on a dried surface
PG6 PG5 Gum PU sponge (DJ) [1]	Comparing the best result obtained with hydrogels with the traditional dry cleaning method	1a	PG6 (90 s + 90 s) + PG5 Gum on a dried surface
		1b	Mechanical removal. Sponges previously washed in demineralized water
PG6 PG5 Gum PU sponge (DJ) [1]	Evaluating the boost in efficacy by combining the two methods	5a	PU sponge (DJ) [1] + PG6 (90 s) + PG5 Gum on a dried surface
		5b	PU sponge (DJ) [1] + PG6 (120 s)
		5c	PU sponge (DJ) [1] + PG5 Gum on a dried surface

[1] DJ = Deffner & Johann®

2.2.3. Assessment of Cleaning Results

Colorimetric analyses were used to assess the efficacy of the cleaning methods in terms of removal of the dirt layer. Three replicas for each measurement were acquired. A Konica Minolta CM-700d colorimeter (Konica Minolta, Osijek, Croatia), with a range of measurement of 400–700 nm, step 10 nm, measurement field of 3–8 mm, d/8 geometry, standard D65 illumination and standard 10° observer was used. The measures were expressed in L*, a* and b* colour space coordinates CIE 1976 and in cylindrical space CIELCH. The specular component included (SCI) data, which allows obtaining results closer to the human eye sensitivity to colours was used. ΔE was calculated using the ΔE_{00}, starting from the colorimetric coordinates of samples before soling and after cleaning operations [18].

Optical microscopy was used on sponges and gels after use, to verify the presence of grains of pigment and thus to evaluate the invasiveness of each test method. Besides this, optical microscopy was carried out to monitor the effects of treatments on the surfaces before and after treatments. The equipment used in this phase was an OLYMPUS SZ X10 (Olympus Corporation, Shinjuku, Tokyo, Japan), interfaced with a PC through a digital camera OLYMPUS Color View I. For capturing and processing the images, analySIS Five software was used.

In addition to this, we documented eventual changes in morphology of the surface, by reflectance transformation imaging (RTI) both before soiling and after the treatments. The RTI technique, based on computational photography, enables the interactive relighting of a subject from any direction, and it is normally used on small areas to emphasize tiny aspects of the surface [19,20]. The samples selected were the ones with the calcite layer only, which are more sensitive to water-based treatments, and with Egyptian Blue and Green ones, for their grain size (respectively the type A, B and C).

Referring to the preliminary measures acquired on the case study, necessary to better calibrate the conservation treatments, we carried out conductivity and superficial pH measurements; the first one has been acquired to work in isotonic conditions with the original painted surface and the second to avoid ionizing action of the cleaning solution [21]. A 2 mm thick pad of agarose (4% in demineralized water) was applied on the object surface for 120 s after having removed the main layer of dust. Conductivity measurement was performed with a LAQUAtwin conductivity meter EC-22 range (Horiba, Kyoto, Japan) and pH analysis was performed with a Hanna Instrument HI 981037 Skin and Scalp pH Tester (Hanna Instruments, Woonsocket, RI, USA).

3. Results and Discussion

3.1. Painting Materials Characterization

Figure 5 shows the OM images of the two sections sampled from the wig (Figure 5a, sample A) and from the body (Figure 5b,c sample B) of the sculpture. The main painting layer scheme described in Figure 2 is visible. In particular, from the bottom to the top of the wig sample a green-white-red-brown sequence is clearly distinguishable (Figures 5a and S1). In this case the first white preparation layer is not recognizable because it was not included in the sampling procedure. Regarding the sample from the body (Figure 5b with a detail in Figure 5c) the first white preparation layer can be seen in the bottom part of the stratigraphy. In this case the sequence is white-blue-white-green-brown. The brown layer seems to be thicker in the body compared to the wig. By means of VIL, SEM-EDX, FTIR, Py-GC/MS and FT-IR all the principal painting layers observed in OM images were identified. Micro-PIXE was used to obtain additional information on minor and trace elements.

FTIR and micro-PIXE spectral data as well as the SEM-EDX results (elemental analysis) are provided as Supplementary Material.

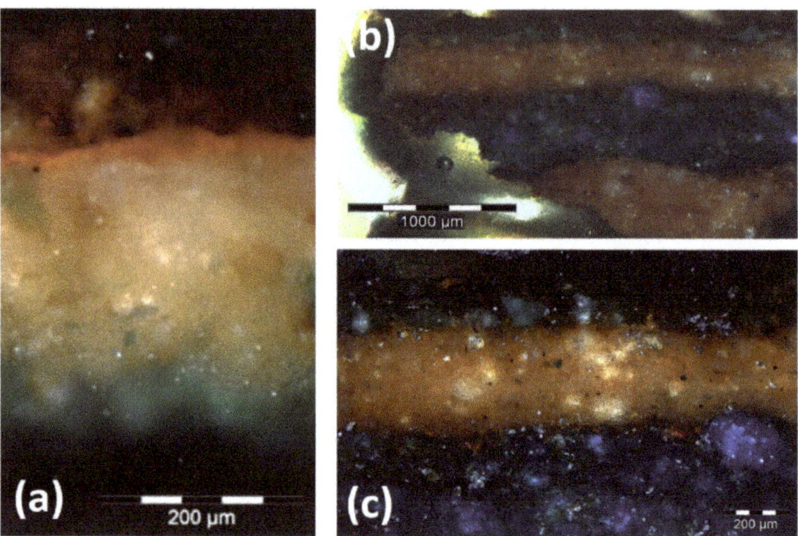

Figure 5. Optical Microscopy images (OM) of the two samples taken from the wig (**a**), sample A, and the body (**b,c**), sample B, of the sculpture; (**c**) is a magnification of a portion of sample B.

3.1.1. Blue and Green Pigments

For what concerns the blue pigment, the preliminary analysis by means of VIL was useful to identify it as Egyptian blue. In Figure 6b a relevant image of VIL compared with photograph is shown. The luminescence was observed only in the body part of the sculpture and not in the head. VIL results were confirmed by means of SEM-EDX on the sample B where a thick blue layer is present. The stoichiometry obtained from homogeneous blue crystals by means of SEM-EDX (Figure S2) is very close to that of cuprorivaite ($CaCuSi_4O_{10}$). In the blue layer, a significant amount of quartz, partially bonded together with a glass, is present, coherent with available literature [22].

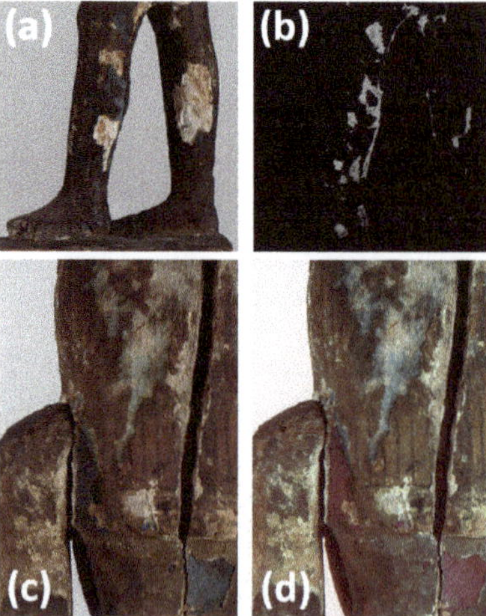

Figure 6. Photographs of part of legs (**a**) and of back/wig; (**c**) VIL image of the same part of the legs (**b**) in which the luminescent areas are made of Egyptian blue; (**d**) IRFC image of the back/wig in which green pigment (turquoise color) is more evident.

The green pigment was observed both in the wig as the first layer over the white preparation layer, and as a final layer under the brown uppermost layer in the body. In Figure 6d the IRFC image of part of the back and wig shows a turquoise tone where the thin green layer is located. Instead, in the same picture, the Egyptian blue shows a red tone. By means of SEM-EDX analysis it was possible to confirm that the pigment is Egyptian Green or green frit. The use of Egyptian Green in antiquity seems to be confined to Egyptian territory, with first evidence in the last part of the third millennium BCE [23]. In terms of microstructure, the green frit consists of glass, from which wollastonite ($CaSiO_3$) and a high temperature polymorph of silica have crystallized, together with partially reacted quartz particles [22]. In Figure 7 is shown a SEM-BS image of a green crystal observed in sample B in which the two-phase microstructure is clear. In particular, by means of SEM-EDX (Figure S3) it was detected that the light grey part of the crystal is made of wollastonite with the addition of sodium and copper, whereas the dark grey part is a silica-rich amorphous phase with calcium, sodium and copper. In average, in the two samples, the green portion has a lower copper content and a higher sodium content than blue portion. These differences are imputable to the different production processes as described by [22,24]. Preliminary results by means of micro-PIXE (Figure S4) have shown some differences in trace elements composition, in particular in potassium content, for the green sectors from sample A and sample B (i.e., the green pigment that form the first buried layer in the wig and the green pigment below the brown layer in the body), even though counting statistic is low and further analyses are necessary to confirm the observation and to correlate it to the chronology of the layer sequence.

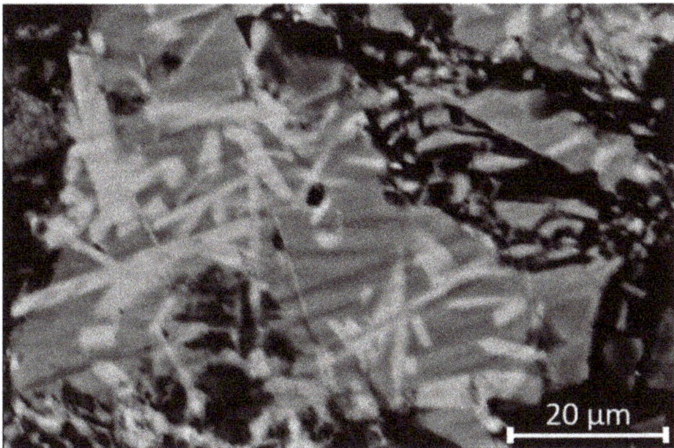

Figure 7. SEM-BS image of a green portion observed in sample B where it is possible to observe two-phase system (light gray is wollastonite, dark gray is high temperature polymorphs of silica).

3.1.2. The brown Pigment Layer

For what concerns the development of a suitable cleaning procedure, the last layer, i.e., the surface brown pigment, is indeed the most important to be studied (Figure 8). This layer covered most of the polychromies of the sculpture, raising a question about its removal. In that sense, the evaluation of its pertinence to the original artistic technique would have been fundamental to consider its eventual removal. Analyses have shown that it is composed by an organic material in which are included many different mineral crystals. FTIR analysis was not able to distinguish any features due to the strong presence of oxalates.

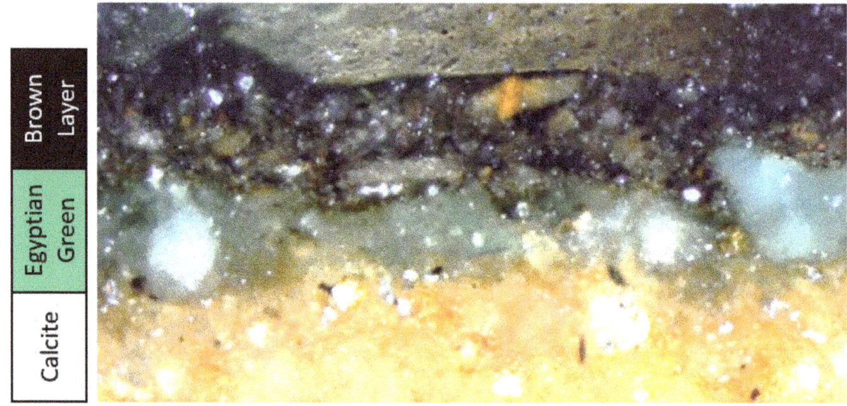

Figure 8. OM images of a details of the stratigraphy of sample B in which is highlighted the complexity of the brown layer over the Egyptian green.

In order to understand the nature of the organic components of the brown material, a pyrolysis-gas chromatography/mass spectrometry analysis was carried out, derivatizing the sample with TMAH. The analysis of the pyrogram allowed to identify a series of marker compounds of substances compatible with materials of natural origin used in ancient Egypt, while no organic materials of synthetic origin were identified. Table 3 contains the list of all peaks for which it was possible to make a certain assignment. Minimal traces of protein markers were also identified, not indicated in the table as they are not significant to clarify the nature of the sample. The Py-GC/MS curve is shown in Figure 9.

Table 3. Marker compounds of organic materials identified by Py-GC/MS in the brown surface layer of the sculpture.

Peak n.	Retention Time [min]	Assignment
1	5.95	1,2,3-Trimethoxypropane
2	8.08	2-Butendioic acid dimethyl ester
3	8.27	2-Butandioic acid dimethyl ester
4	9.32	Benzoic acid methyl ester
5	11.71	2-Methoxybutendioic acid dimethyl ester
6	12.15	Permethylated 3-deoxypentenoic acid methyl ester
7	12.40	Permethylated 3-deoxypentenoic acid methyl ester
8	13.16	Permethylated 3,6-deoxyhexenoic acid methyl ester
9	13.32	1,2,4-Trimethoxybenzene
10	13.40	Permethylated 3,6-deoxyhexenoic acid methyl ester
11	13.42	4-Methoxybenzoic acid methyl ester
12	13.87	1,2,3-Propaentricarboxylic acid trimethyl ester
13	14.25	Octanedioic acid dimethyl ester
14	14.69	Permethylated 3-deoxyhexenoic acid methyl ester
15	15.17	Permethylated 3-deoxyhexenoic acid methyl ester
16	15.49	Nonanedioic acid dimethyl ester
17	16.08	3,4-Dimethoxybenzoic acid methyl ester
18	16.39	2,3,4,6-Tetra-O-methyl-D-gluconic acid δ-lactone
19	17.51	Tetradecanoic acid methyl ester
20	17.56	3,4,5-Trimethoxybenzoic acid methyl ester
21	18.67	1,2,3-Benzentricarboxylic acid trimethyl ester
22	18.86	1,2,4-Benzentricarboxylic acid trimethyl ester
23	19.39	9-Hexadecenoic acid methyl ester
24	19.62	Hexadecanoic acid methyl ester

Table 3. *Cont.*

Peak n.	Retention Time [min]	Assignment
25	21.30	9-Octadecenoic acid methyl ester
26	21.54	Octadecanoic acid methyl ester
27	23.27	Eicosanoic acid methyl ester
28	24.89	Docosanoic acid methyl ester
29	25.65	21-Methyldocosanoic acid methyl ester
30	26.15	Heptacosane
31	26.40	Tetracosanoic acid methyl ester
32	26.86	Octacosane
33	27.09	Squalene
34	27.59	Nonacosane
35	27.86	Hexacosanoic acid methyl ester
36	28.38	Tricontane
37	29.36	3-Methoxycholest-5-ene
38	29.65	Octacosanoic acid methyl ester

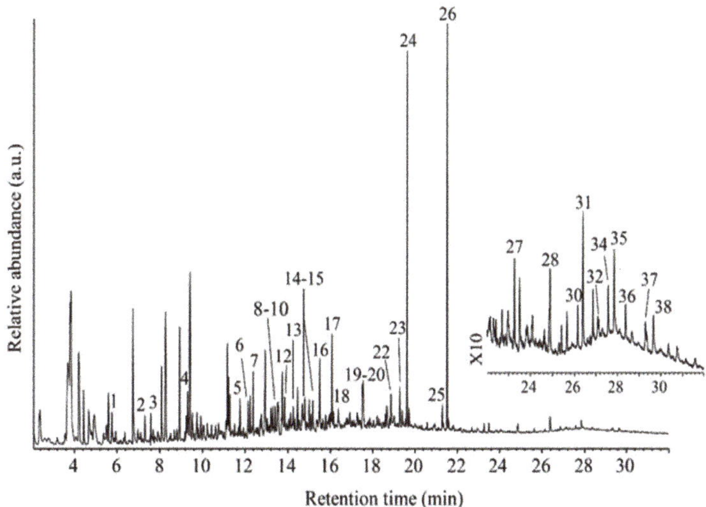

Figure 9. Pyrogram of the surface paint layer of the sculpture. For assignments see Table 3.

The two most intense signals are attributed to palmitic (peak n. 24) and stearic acid (n. 26), detected in the form of methyl esters. Longer chain saturated fatty acids have also been identified, with a number of carbon atoms from C_{20} to C_{28} (n. 27, 28, 29, 31, 35, 38), and some unsaturated fatty acids (n. 23, 25). These compounds, together with C_{27}–C_{30} linear alkanes (n. 30, 32, 34, 36), are markers of lipidic substances such as natural waxes. However, the specific markers of the main natural waxes (e.g., 15-hydroxyhexadecanoic acid for beeswax) are missing. The presence of cholesterol (n. 37) could be due to the use of an animal wax or animal fats, not better identified and probably mixed with other fatty substances [25,26]. These same compounds are also present in human sebum and could be due to the manipulation of the object, a hypothesis also supported by the presence of squalene, a marker compound present in fingermarks [27]. Moreover, the peaks assigned to glycerol (n. 1) and dicarboxylic acids with eight and nine carbon atoms (here detected as dimethyl esters, n. 13 and 16) can be related to glycerol-based lipids of oils, fats or body lipids of microorganism that might be present in the sample.

The other compounds identified by Py-GC/MS belong to two different chemical classes, saccharides and aromatic compounds such as phenols and hydroxyaromatic acids.

As for saccharides, markers derived from arabinose (n. 6, 7), rhamnose (n. 8, 10) and galactose (n. 14, 15) have been detected, as expected for gum Arabic [28]. Gum Arabic was extensively used as binding medium in ancient Egypt, therefore its finding on the surface of the sculpture is consistent with the presence of a paint layer [23,29]. Furthermore, yellowing of gum Arabic due to aging has been associated with the darkening of paints containing Egyptian blue, which in many ancient artefacts appear brownish green or almost black [30]. This phenomenon, observed on several objects decorated with Egyptian blue paints, could be co-responsible for the current visual aspect of the sculpture.

Some of the aromatic compounds identified (n. 17, 20) could be attributed to a contamination due to the wood material of the sculpture [31]. However, the building-block compounds of lignin (i.e., methyl, ethyl, *n*-propyl and vinyl guaiacols) are absent, so it appears more likely that their origin is different. The same compounds were identified in Egyptian mummification balms and it was hypothesized that they are oxidation products of balsamic resins secreted by plants of the Umbelliferae family [32]. Hydroxyaromatic acids are also markers of humic acids and tannin-derived materials (n. 2, 3, 4, 5, 9, 11, 17, 20) [33,34]. The latter are particularly interesting because of their brown colour which tends to black when combined with iron [35]. These compounds could also contribute to the dark colour of the superficial layer of the sculpture.

To investigate in deep the brown layer, a petrographic study of the mineral grains included in the gum was carried out to verify its compatibility with an Egyptian territory origin. In particular, a fine sand formed by clasts smaller than 100 microns is present (Figure 10).

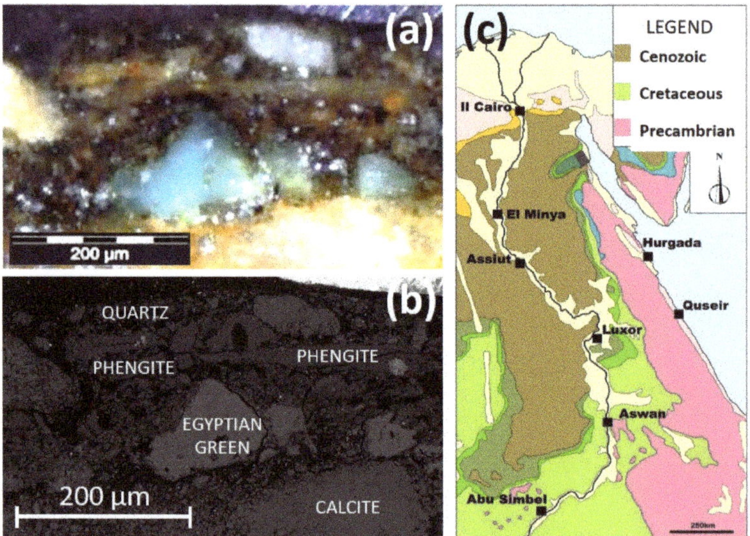

Figure 10. OM (**a**) and SEM-BS (**b**) images of the surface layer that is composed by an organic material in which different minerals are dispersed; simplified geological map of Egypt (**c**).

Numerous mineral particles were observed in this layer consisting mainly of quartz, calcite and white mica. In accessory quantities there are also biotite, pyroxene, iron oxides, apatite and sulphides. Respect to the provenance study, among the various minerals found, the most interesting was potassium white mica. It constantly shows a phengitic composition, in the sense that it shows an enrichment in Si and a reduction in Al, compared to the theoretical formula of muscovite [36]. In the literature, phengite is commonly associated with metamorphic rocks that formed under conditions of high pressure, in the subduction zones [37]. This therefore allows to confirm the compatibility with Egyptian territory and to constrain the area of origin of the raw material used by the Ancient Egyptians, as part of the covering layer of the statue in question. By observing a simplified geological map of Egypt, the territory can be divided into three main geological units (see Figure 10c). In the central-northern sector of Egypt, sedimentary rocks of carbonate origin from the Cenozoic age mainly crop out [38]. The presence of abundant silicate clasts in the analysed sample makes it possible to exclude with good approximation that the material used comes from this sector of Egypt and, in particular, from the delta area of the Nile. In the southern sector of Egypt, on the other hand, sedimentary rocks of the Cretaceous age referred to the Nubian Sandstone Formation occur [39]. Even the outcrop area of these sandstones can be excluded as the area of origin of the raw material, as the Nubian sandstones are extremely pure and almost exclusively made up of quartz clasts. Finally, the eastern sector of Egypt is characterized by the presence of very ancient crystalline rocks (pre-Cambrian in age, corresponding over to 500 million years), which are called Arab-Nubian shield. To this geological unit belongs both the famous Aswan granites and metamorphic units of continental crust [40]. Granites very rarely contain white mica and therefore it can be excluded that the material comes from the Aswan area. Instead, the different metamorphic units out cropping in the Egyptian Eastern Desert are characterized by metamorphic conditions favourable to the stability of phengitic mica. In particular, the eastern desert is crossed by the Wadi Hammamat, an ancient road link between the Nile and the Red Sea, frequented by the Ancient Egyptians since the fourth dynasty and especially in the Ramesseid era, a period to which the production of the Papyrus of the Mines also dates back [41]. Therefore, based on the mineralogical data collected using the SEM analysis, it is possible to infer that the raw material for the covering layer of the sculpture is compatible with Egyptian territory. Probably it comes from areas of the eastern desert and was transported along the Wadi Hammamat, while other sources such as the Nile delta and the Aswan area are to be excluded.

3.1.3. Other Pigments

The red pigment in the sample A (taken from the wig) was attributed to red ochre because from SEM-EDX analysis it turns out to be rich in iron with minor contents of other elements such as silicon, aluminium, magnesium and potassium (Figure S1). The result was confirmed also by means of micro-PIXE measurements. Red ochre was a very common pigment used starting from the fourth millennium BCE through the Roman period [23].

All the white layers used as preparation, both in sample A and sample B, are made of calcite, another very common material employed in Ancient Egypt starting from the Predynastic Period [23]. No presence of sulphur as main element was observed, excluding the use of gypsum or anhydrite. Moreover, from preliminary micro-PIXE (Figure S4) analysis no particular differences were observed in minor and trace elements (Si, S, Cl, Fe and Cu) in the intermediate and first layers, even though the result is not sufficient to hypothesize a contemporaneity of the two layers.

For what concerns the decorative elements, FT-IR analysis (Figure S5) carried out on a sample from the white belt have shown the presence of huntite, $Mg_3Ca(CO_3)_4$, a carbonate mineral which provides a brighter white than calcite. Its use in Ancient Egypt is documented starting from third millennium BCE [23].

No analyses were considered necessary to understand the artwork for conservation purposes on black and yellow decorations. The black pigment is made probably of charcoal or carbon also considering their strong absorption in IR images, and the yellow pigment was attributed to yellow ochre, in consideration of literature, its hue and morphology [42].

3.2. Cleaning Tests on Mockups

Following the conservation cleaning treatments guidelines, preliminary tests were carried out on mockups (see Table 1) to evaluate the effect of each selected method with the aim of defining the safest and most efficient cleaning protocol to treat the original surface.

The first cleaning tests were performed using the two selected cleaning methods alone, i.e., PU sponges and highly retentive hydrogels. In particular, we focused in finding the best combination of application length and number of applications for PG6. Then, we evaluate the effectiveness of PG5 Gum to finalize a cleaning procedure performed with PG6. Results obtained using the two cleaning methods alone were then evaluated. Afterwards, in order to take advantage of the strengths of the tested methods and to minimize their weaknesses, we combined the dry and the water-based cleaning treatments. The characterization of samples before and after cleaning tests allowed to evaluate the performances of the materials and to determine the failure point of each treatment.

Several tests were carried out using PG6 to define a time range to work safely and efficiently on the surface to be cleaned, i.e., the maximum and minimum length of application were defined. Thanks to the OM, we observed that after the application of a PG6 for 180 s, a partial alteration of the substrate took place, as testified either by the presence of bigger grains on the gel's surface in contact with the substrate to be cleaned or by colour changes in the applied gel. On the other hand, applications shorter than 90 s did not result an effective cleaning of the surface.

Moreover, we compared the effect of a single long application with two subsequent shorter applications, having the same or higher overall contact time. For instance, we noticed that a single 2-min-long application allowed to obtain good cleaning results but caused a partial migration of the pigment from the surface. On the other hand, a two steps application provided comparable results, without changes in the original materials, granting higher control on the cleaning action. Overall, the most promising results were obtained with a two-step application of PG6 (3a test, 90 s + 90 s), although some residues of dirt were still present on the surface, requiring a localized refining of the cleaning.

To that aim, PG5 Gums were tested both on dry (after complete evaporation of the water released by PG6) and wet (immediately after the application of PG6) surfaces. PG5 Gums performed in a satisfying way for the localized removal of dirt residues. It is worth noting that, during application, the gum should be gently handled (not squeezed) to prevent uncontrolled release of water. Overall, the best results were obtained by the application of PG5 Gum over the dry surface, which withstands a mechanical action, even if gentle, better than wet and softened materials.

The single methods (PU sponges and the best PG systems combination) showed a good potential for the cleaning of water-sensitive surfaces, even if both displayed some limitations. In particular, PU sponges efficiently removed the dust layer, but at the expense of the integrity of the original paint layer. In fact, several coloured particles were detected on the surface of the sponges, especially when applied on mockups prepared using pigments with larger grains (see Figure 11a). On the contrary, highly retentive hydrogels did not interfere with the original material, i.e., no coloured particles were detected on gel's surfaces (see Figure 11b), but the system only partially removed the dirt layer, providing an incomplete cleaning effect.

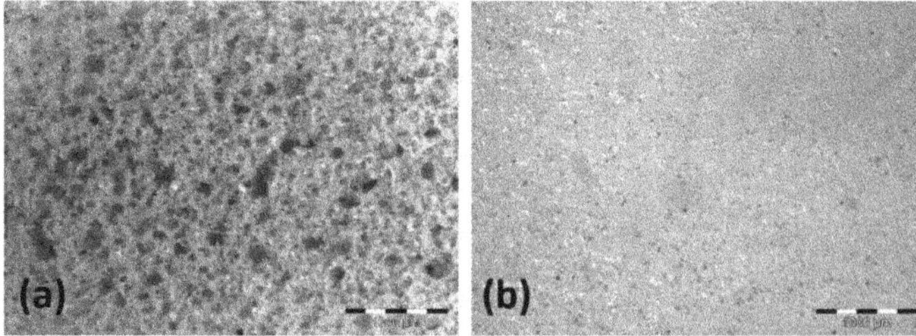

Figure 11. (**a**) OM image (25×) of the PU sponge after cleaning at the interface of green sample (set n.1); (**b**) OM image (25×) of the PG6 after cleaning at the interface of blue sample (set n.2).

For this reason, as indicated in Table 2, we tested different combination of the two methods, to define an effective cleaning procedure (tests' series 5). The best results were obtained with a preliminary gentle dry-cleaning step that provided a partial removal of the dirt without altering the original surface. The second step, carried out with PG6 (90 s) removed the dirt layer left on the surface without damaging the treated material. PG5 was then applied on dried samples, refining the cleaning with a localized action completely respectful of the original material (5a set, Figure 12).

Figure 12. 5a set after cleaning tests using best conditions. Not soiled area on the top.

To confirm the optical evaluation and the OM results, colorimetric measurements were carried out on this set of samples before soiling and after cleaning. The obtained ΔE_{00} span between a maximum of 8.9 and a minimum of 1.3, with higher values corresponding to samples B and C (Egyptian Blue and Green without the top brown layer), because of their superficial roughness, which hampered a homogenous cleaning, and lower values measured on samples A and F (preparation layer only and ochre with yellow stripes). Nevertheless, considering the complexity of the prepared samples, the colorimetric changes measured were deemed acceptable and the overall cleaning results were considered satisfactory. Moreover, RTI analysis did not detect any significant morphological change on the surface after the combined application of PU sponges or the innovative hydrogels (Figure 13), confirming the safeness of the used cleaning procedure.

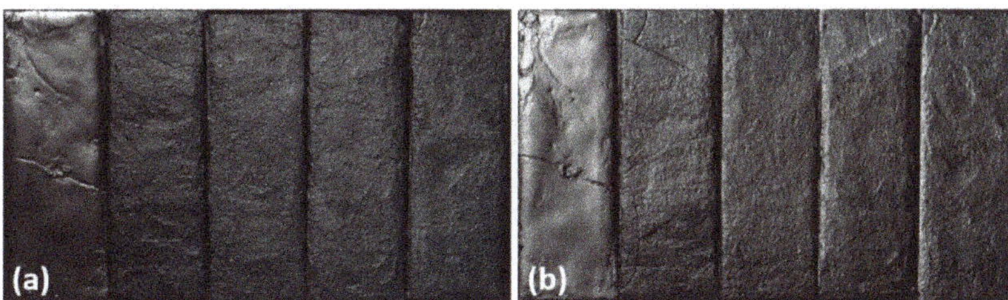

Figure 13. RTI documentation before (**a**) and after (**b**) cleaning tests.

3.3. Cleaning Treatment of the Egyptian Statuette

A careful analysis of the pictorial surface revealed problems of stability for the entire stratigraphy, which showed a loss in the adhesion at multiple stratigraphic levels often combined with partial detachments of the material at the underlying interface. Moreover, due to the loss of cohesion of the Egyptian Blue layer belonging to the first pictorial level, the sculpture showed several cracks within this layer, with a consequent embrittlement of the entire stratigraphy. Therefore, it was decided to consolidate the pictorial layers before the cleaning treatment, by injecting locally, only where needed, a solution of hydroxyl propyl cellulose (Klucel G®, CTS Europe®, Milan, Italy, 2% in ethanol), which was deemed compatible with the original materials.

After that, based on the results obtained on mockups, we selected the best procedure for the superficial cleaning of the Egyptian statuette. The measure of pH (6.62) and conductivity (3 µS/cm) suggested not to consider necessary to use a buffer solution, so as to avoid any need of rinsing the surface with a subsequent application of water-loaded gels to remove buffer residues.

The tests conducted on mockups were fundamental for the cleaning of the artefact. However, as expected, some adjustments have been performed during cleaning operations, especially where a thicker layer of dirt was present. In those areas, a 2-min-long single application of PG6 was needed to remove most of the soil. The finishing of the cleaning was carried out with PG5 Gum, which were deemed particularly useful in areas with major undercuts.

Overall, the selected cleaning procedure of the artefact increased the perception of the painted decoration, leading to a general matting of the surface that results in a clearer tone of the brown layer, according to its supposed original appearance (Figure 14).

Figure 14. The artefact after the cleaning treatments.

4. Conclusions

The characterization of the original materials collected from the surface of the statue confirmed the use of a colours' palette typical of the Ancient Egyptian production and a complex layering of the polychromies. For what concerns the white pigment, calcite was used for the preparation layers and huntite for the belt decoration. Red ochre was employed for the wig and probably for the base of the sculpture. Moreover, Egyptian Blue was found in samples taken for the body, whereas Egyptian Green was used to colour part of the body and the wig. Interestingly, a brown pigment was used to cover the whole sculpture. Analyses have shown that it is composed by a fine sand and gum Arabic as binder. Based on the mineralogical data collected, it is possible to infer that the raw material for the covering layer of the sculpture is compatible with Egyptian territory. Probably it comes from areas of the eastern desert, while other sources such as the Nile delta and the Aswan area are to be excluded. The dark colour of this layer might be due to several reasons: we hypothesize that the original brown colour was darken by the alteration of the gum Arabic used as binder, and by the addition of humic acids and tannin-derived materials that could be related to the destination of use of the sculpture, possibly a funerary one.

The cleaning of ancient Egyptian artefacts is still an open problem, because it often implies the removal of overlapped materials and superficial dirt from hydrophilic, porous and extremely delicate surfaces, which often do not feature any finishing layer. For those reasons, the dry cleaning is often preferred to solvents or water-based cleaning methods. However, this specific case study had further challenges: its three-dimensionality, the fragility of the original materials, the complex layering and the irregular morphology of the surfaces complicated the cleaning process, and the chromatic similarity between the non-homogeneous dirt layer and the underlying brown pigment layer below needed an even more careful monitoring of the cleaning operations.

The application of PU sponges on mockups allowed for the almost complete removal of soil, but at the expense of the integrity of the original paint layer. In fact, several pigment grains were removed for the surface together with the dirt layer. The best results in terms of cleaning effectiveness and non-invasiveness to the original surfaces have been obtained by a gentle action using PU sponges followed by the application of highly retentive polyvinyl alcohol-based gels, namely PG6 and PG5 gums. The preliminary application of PU sponges allowed for the partially removal of the soil without altering the original surface, while the gels permitted a gradual and controlled action at the interface without removing pigments' grains. Following the promising results obtained on mockups, the ancient Egyptian statuette was cleaned successfully and safely.

To summarize, thanks to this study, we had the chance of collecting new insights about the chemical composition of the artefact, which can be fundamental for archaeologists and art historians. Moreover, it was demonstrated that, when confined in highly retentive gels, water-based systems can be safely used for the cleaning of hydrophilic surfaces. Future perspective may involve additional testing of these flexible and elastic hydrogels on other artistic surfaces that are highly reactive to aqueous-based treatments, with the aim of expanding the palette of available tools for conservators working on fragile, sensitive and delicate works of art, improving the results that can obtained with the sole traditional dry-cleaning methodologies.

Supplementary Materials: The following are available online at https://www.mdpi.com/article/10.3390/coatings11111335/s1, Figure S1: SEM-EDX maps of the main elements in the sample A. The first image in gray scale is the SEM-BSE image, Figure S2: Elemental analysis (weight %) by means of SEM-EDX of three representative blue grains. The blue squares are the areas of analysis. On top right are shown the optical images of the grains, Figure S3: Elemental analysis (weight %) by means of SEM-EDX of a representative green grain. The green squares are the areas of analysis. In the center is shown the optical images of the grain, Figure S4: Semi-quantitative elemental analysis by means of PIXE of different green (top) and white preparation (bottom) layers, Figure S5: FT-IR analysis carried out on a sample from the white belt has shown the presence of huntite.

Author Contributions: Conceptualization, N.M., P.B. (Paola Buscaglia) and A.L.G.; methodology, P.B. (Paola Buscaglia) and A.L.G.; investigation, A.B., A.R., D.S., L.G., M.N., N.M., G.P. and P.B. (Piero Baglioni); data curation, P.B. (Paola Buscaglia), A.B., A.L.G., D.S., L.G., M.N., N.M.; supervision, M.B., P.G., S.A., P.B. (Paola Buscaglia) and A.L.G.; writing—original draft preparation, N.M., A.B., A.L.G., D.S., and P.B. (Paola Buscaglia); writing—review and editing, A.L.G., G.P., P.B. (Piero Baglioni) and P.B. (Paola Buscaglia); visualization, P.B. (Paola Buscaglia) and A.L.G. All authors have read and agreed to the published version of the manuscript.

Funding: The European Union (NANORESTART and APACHE projects, Horizon 2020 research and innovation program under grant agreement No 646063 and 814496, respectively) is gratefully acknowledged for partial financial support.

Acknowledgments: The authors wish to warmly thank Anna Piccirillo (Centro Conservazione e Restauro La Venaria Reale) and Tommaso Poli (Dipartimento di Chimica-Università degli Studi di Torino) for the FT-IR analyses and Daniele Demonte (Centro Conservazione e Restauro La Venaria Reale) for the multiband imaging.

Conflicts of Interest: The authors declare no conflict of interest.

References

1. Baines, J. *Fecundity Figures. Egyptian Personification and the Iconology of a Genre*; Aris & Phillips Ltd.: Warminster, UK, 1985; pp. 317–318, ISBN 0-85668-087-7.
2. Mottica, R. Scultura Lignea in Terra d'Egitto: Analisi Valutativa di una Statuetta del Dio Nilo Conservata al Museo Egizio di Torino. Bachelor's Thesis, Università di Torino, Torino, Italy, 2006.
3. Santoro, S. Il Colore nell'Antico Egitto: Inquadramento e Riflessioni Tecnico-Artistiche e Simboliche sui Colori di Statuetta del dio Nilo, Integrate da Indagini Archeometriche al SEM-EDS. Bachelor's Thesis, Università di Torino, Torino, Italy, 2006.
4. Serrapede, M. I Colori del Dio Nilo: Caratterizzazione delle Stesure Cromatiche di una Statuetta Lignea del Dio per Mezzo di XRF, PIXE e SEM-EDS. Bachelor's Thesis, Università di Torino, Torino, Italy, 2007.
5. Manfredda, N. Conservation Problems of an Artefact with Double Layers of Polychromy: A Wooden Sculpture of New Kingdom from the Museo Egizio of Torino. Master's Thesis, Università di Torino in agreement with Centro Conservazione e Restauro la Venaria Reale, Torino, Italy, 2019.
6. Fabretti, A.; Rossi, F.; Lanzone, R.V. *Regio Museo di Torino. Antichità Egizie*; Stamperia Reale: Torino, Italy, 1882; Volume 1, p. 58.
7. Daudin-Schotte, M.; Bisschoff, M.; Joosten, I.; van Keulen, H.; van den Berg, K.J. Dry cleaning approaches for unvarnished paint surfaces. In *New Insights into the Cleaning of Paintings: Proceedings from the Cleaning 2010 International Conference, Universidad Politécnica de Valencia and Museum Conservation Institute*; Mecklenburg, M.F., Charola, E., Koestler, R.J., Eds.; Smithsonian Contributions to Museum Conservation; Smithsonian Institution Scholarly Press: Washington, DC, USA, 2013; pp. 209–219.
8. Baglioni, M.; Poggi, G.; Chelazzi, D.; Baglioni, P. Advanced materials in cultural heritage conservation. *Molecules* **2021**, *26*, 3967. [CrossRef]
9. Domingues, J.A.L.; Bonelli, N.; Giorgi, R.; Fratini, E.; Gorel, F.; Baglioni, P. Innovative hydrogels based on semi-interpenetrating p(HEMA)/PVP networks for the cleaning of water-sensitive cultural heritage artifacts. *Langmuir* **2013**, *29*, 2746–2755. [CrossRef] [PubMed]
10. Bonelli, N.; Montis, C.; Mirabile, A.; Bertia, D.; Baglioni, P. Restoration of paper artworks with microemulsions confined in hydrogels for safe and efficient removal of adhesive tapes. *Proc. Natl. Acad. Sci. USA* **2018**, *115*, 5932–5937. [CrossRef]
11. Cardaba, I.; Poggi, G.; Baglioni, M.; Chelazzi, D.; Maguregui, I.; Giorgi, R. Assessment of aqueous cleaning of acrylic paints using innovative cryogels. *Microchem. J.* **2020**, *152*, 104311. [CrossRef]
12. Bonelli, N.; Poggi, G.; Chelazzi, D.; Giorgi, R.; Baglioni, P. Poly(vinyl alcohol)/poly(vinyl pyrrolidone) hydrogels for the cleaning of art. *J. Colloid Interface Sci.* **2019**, *536*, 339–348. [CrossRef] [PubMed]
13. Bartoletti, A.; Barker, R.; Chelazzi, D.; Bonelli, N.; Baglioni, P.; Lee, J.; Angelova, L.V.; Ormsby, B. Reviving WHAAM! a comparative evaluation of cleaning systems for the conservation treatment of Roy Lichtenstein's iconic painting. *Herit. Sci.* **2020**, *8*, 9. [CrossRef]
14. Mastrangelo, R.; Chelazzi, D.; Poggi, G.; Fratini, E.; Pensabene Buemi, L.; Petruzzelis, M.L.; Baglioni, P. Twin-chain polymer hydrogels based on poly(vinyl alcohol) as new advanced tool for the cleaning of modern and contemporary art. *Proc. Natl. Acad. Sci. USA* **2020**, *117*, 7011–7020. [CrossRef]
15. Vigorelli, L.; Re, A.; Guidorzi, L.; Cavaleri, T.; Buscaglia, P.; Nervo, M.; Del Vesco, P.; Borla, M.; Grassini, S.; Lo Giudice, A. Multi-analytical approach for the study of an ancient Egyptian wooden statuette from the collection of Museo Egizio of Torino. *Acta Imeko* **2021**, in press.
16. Petrakakis, K.; Dietrich, H. MINSORT: A program for the processing and archivation of microprobe analysis of silicate and oxide minerals. *Neues Jb. Miner. Abh.* **1985**, *8*, 379–384.
17. Lo Giudice, A.; Re, A.; Angelici, D.; Corsi, J.; Gariani, G.; Zangirolami, M.; Ziraldo, E. Ion microbeam analysis in cultural heritage: Application to lapis lazuli and ancient coins. *Acta Imeko* **2017**, *6*, 76–81. [CrossRef]
18. Oleari, C. *Misurare il Colore*, 2nd ed.; Hoepli: Milano, Italy, 2008; pp. 222–224, ISBN 8-820-34126-3.
19. Piquette, K.E. Reflectance transformation imaging (RTI) and Ancient Egyptian material culture. *Damqatum CEHAO Newsl.* **2011**, *7*, 16–20.
20. Serotta, A. An investigation of tool marks on Ancient Egyptian hard stone sculpture: Preliminary report. In *Metropolitan Museum Studies in Art, Science, and Technology*; Centeno, S.A., Kennedy, N.W., Manuels, M., Schorsch, D., Stone, R.E., Sun, Z.J., Wypyski, M.T., Eds.; Metropolitan Museum of Art: New York, NY, USA, 2014; Volume 2, pp. 197–201, ISBN 0-300-20439-6.
21. Wolbers, R. *Cleaning Painted Surfaces: Aqueous Methods*; Archetype Publications: London, UK, 2000; ISBN 978-1-873132-36-4.
22. Hatton, G.D.; Shortland, A.J.; Tite, M.S. The production technology of Egyptian blue and green frits from second millennium BC Egypt and Mesopotamia. *J. Archaeol. Sci.* **2008**, *35*, 1591–1604. [CrossRef]
23. Scott, D.A. A review of ancient Egyptian pigments and cosmetics. *Stud. Conserv* **2016**, *61*, 185–202. [CrossRef]
24. Pagès-Camagna, S.; Colinart, S. The Egyptian green pigment: Its manufacturing process and links to Egyptian blue. *Archaeometry* **2003**, *45*, 637–658. [CrossRef]

25. Asperger, A.; Engewald, W.; Fabian, G. Advances in the analysis of natural waxes provided by thermally assisted hydrolysis and methylation (THM) in combination with GC/MS. *J. Anal. Appl. Pyrolysis* **1999**, *52*, 51–63. [CrossRef]

26. Asperger, A.; Engewald, W.; Fabian, G. Thermally assisted hydrolysis and methylation—A simple and rapid online derivatization method for the gas chromatographic analysis of natural waxes. *J. Anal. Appl. Pyrolysis* **2001**, *61*, 91–109. [CrossRef]

27. Girod, A.; Weyermann, C. Lipid composition of fingermark residue and donor classification using GC/MS. *Forensic Sci. Int.* **2014**, *238*, 68–82. [CrossRef] [PubMed]

28. Riedo, C.; Scalarone, D.; Chiantore, O. Advances in identification of plant gums in cultural heritage by thermally assisted hydrolysis and methylation. *Anal. Bioanal. Chem.* **2010**, *396*, 1559–1569. [CrossRef] [PubMed]

29. Riedo, C.; Scalarone, D.; Chiantore, O. Multivariate analysis of pyrolysis-GC/MS data for identification of polysaccharide binding media. *Anal. Methods* **2013**, *5*, 4060–4067. [CrossRef]

30. Daniels, V.; Stacey, R.; Middleton, A. The blackening of paint containing Egyptian blue. *Stud. Conserv.* **2004**, *49*, 217–230. [CrossRef]

31. Challinor, J.M. Characterisation of wood by pyrolysis derivatisation-gas chromatography/mass spectrometry. *J. Anal. Appl. Pyrolysis* **1995**, *35*, 93–107. [CrossRef]

32. Buckley, S.A.; Evershed, R.P. Organic chemistry of embalming agents in Pharaonic and Graeco-Roman mummies. *Nature* **2001**, *413*, 837–841. [CrossRef] [PubMed]

33. Fabbri, D.; Chiavari, G.; Galle, G.C. Characterization of soil humin by pyrolysis(/methylation)-gas chromatography/mass spectrometry: Structural relationships with humic acids. *J. Anal. Appl. Pyrolysis* **1996**, *37*, 161–172. [CrossRef]

34. Kaal, J.; Nierop, K.G.J.; Kraal, P.; Preston, C.M. A first step towards identification of tannin-derived black carbon: Conventional pyrolysis (Py-GC-MS) and thermally assisted hydrolysis and methylation (THM-GC-MS) of charred condensed tannins. *Org. Geochem.* **2012**, *47*, 99–108. [CrossRef]

35. Chiavari, G.; Montalbani, S.; Prati, S.; Keheyan, Y.; Baroni, S. Application of analytical pyrolysis for the characterisation of old inks. *J. Anal. Appl. Pyrolysis* **2007**, *80*, 400–405. [CrossRef]

36. Guidotti, C.V. Micas in metamorphic rocks. *Rev. Mineral.* **1984**, *13*, 357–467.

37. Massone, H.J.; Schreyer, W. Phengite geobarometry based on the limiting assemblage with K-feldspar, phlogopite and quartz. *Contrib. Mineral. Petrol.* **1987**, *96*, 212–224. [CrossRef]

38. De Putter, T.; Karlshausen, C. *Les Pierres Utilisées dans la Sculpture et L'architecture de l'Ègypte Pharaonique: Guide Pratique Illustré*; Connaissance de l'Égypte Ancienne: Bruxelles, Belgium, 1992; ISBN 2-87268-003-9.

39. Klitzsch, E.; Harms, J.C.; Lejal-Nicol, A.; List, F.K. Major subdivisions and depositional environments of Nubia strata, Southwestern Egypt. *Am. Assoc. Pet. Geol. Bull.* **1979**, *63*, 967–974. [CrossRef]

40. Finger, F.; Dörr, W.; Gerdes, A.; Gharib, M.; Dawoud, M. U-Pb zircon ages and geochemical data for the Monumental Granite and other granitoid rocks from Aswan, Egypt: Implications for the geological evolution of the western margin of the Arabian Nubian Shield. *Mineral. Petrol.* **2008**, *93*, 153–183. [CrossRef]

41. Borghi, A.; Vaggelli, G.; D'Amicone, E.; Fiora, L.; Maschali, O.; Shalaby, B.; Vigna, L. Bekhen stone artifacts in the Egyptian Antiquity Museum of Turin (Italy): A mineropetrographic study. In Proceedings of the 2nd International Conference on the Geology of the Tethys, Cairo University, Cairo, Egypt, 19–22 March 2007.

42. Lee, L.; Quirke, S. Painting materials. In *Ancient Egyptian Materials and Technology*; Nicholson, P.T., Shaw, I., Eds.; Cambridge University Press: Cambridge, UK, 2000; pp. 104–120, ISBN 0-521-12098-5.

Article

The Life of a Painting as Traced by Technical Analysis: Original Materials and Posthumous Alterations in Édouard Manet's *Woman in Striped Dress*

Federica Pozzi [1,*,†], **Silvia A. Centeno** [1], **Federico Carò** [1], **Gillian McMillan** [2], **Lena Stringari** [2] **and Vivien Greene** [3]

[1] Department of Scientific Research, The Metropolitan Museum of Art, 1000 Fifth Avenue, New York, NY 10028, USA; silvia.centeno@metmuseum.org (S.A.C.); federico.caro@metmuseum.org (F.C.)

[2] Conservation Department, Solomon R. Guggenheim Museum, 1071 Fifth Avenue, New York, NY 10128, USA; gmcmillan@guggenheim.org (G.M.); lstringari@guggenheim.org (L.S.)

[3] Curatorial Department, Solomon R. Guggenheim Museum, 1071 Fifth Avenue, New York, NY 10128, USA; vgreene@guggenheim.org

* Correspondence: federica.pozzi@centrorestaurovenaria.it

† Current address: Center for Conservation and Restoration of Cultural Heritage "La Venaria Reale", Via XX Settembre 18, 10078 Venaria Reale, TO, Italy.

Abstract: Among the holdings of the Solomon R. Guggenheim Foundation, New York, is a large-scale portrait by Édouard Manet that remained apparently unfinished upon the artist's death, in April 1883. This work, now known as *Woman in Striped Dress*, belongs to Manet's late artistic production and dates from around 1877 to 1880. A collaborative endeavor entailing archival research and scientific analysis revealed that the composition had suffered dramatic alterations prior to its arrival at the museum in 1965 as an extended loan, likely carried out to "finish" the picture in order that it would be marketable and to increase its sale value. Among the main changes explored in this technical study are the reductions in the canvas size and subsequent varnishing campaigns. Furthermore, along with a detailed characterization of the original materials present in the ground and paint layers, this research contributed to the identification of posthumous retouchings, possibly executed concurrently with trimming the canvas along both sides and at the top edge. The investigation was instrumental in devising an appropriate treatment to remove the discolored varnish and select areas of retouching, which obscured significant details of the composition and Manet's delicate brushwork.

Keywords: Manet; painting; portrait; woman; 19th century; pigments; extenders; varnish; original materials; later alterations

Citation: Pozzi, F.; Centeno, S.A.; Carò, F.; McMillan, G.; Stringari, L.; Greene, V. The Life of a Painting as Traced by Technical Analysis: Original Materials and Posthumous Alterations in Édouard Manet's *Woman in Striped Dress*. *Coatings* **2021**, *11*, 1334. https://doi.org/10.3390/coatings11111334

Academic Editor: Maurizio Licchelli

Received: 23 September 2021
Accepted: 26 October 2021
Published: 30 October 2021

Publisher's Note: MDPI stays neutral with regard to jurisdictional claims in published maps and institutional affiliations.

1. Introduction

When Édouard Manet died in April 1883, more than one hundred paintings in various stages of completion were left in his studio in Paris. Among these was a large-scale picture of a woman, possibly unfinished, now known as *Woman in Striped Dress* (Figure 1) [1,2]. Belonging to the artist's late oeuvre and dating from around 1877 to 1880, this painting is housed in the Thannhauser collection at the Solomon R. Guggenheim Museum, New York. It depicts an unknown model, possibly a courtesan dressed up for the occasion, standing before an unidentified background, likely staged inside the artist's studio. This picture resembles many of Manet's late paintings, which often appear to be rapidly executed and characterized by broad, lively brushwork applied in a direct manner. Entire passages of his compositions are typically dominated by just a few paint strokes, in the pursuit of an instantaneous translation of visual experience onto the two-dimensional canvas. As the artist told his lifelong friend Antonin Proust: *"There's just one real thing. To get down what one sees at the first shot. When it's there, it's there. When it's not there, one starts over. Everything else is nonsense"* [3]. However, the artist is known to have struggled between perfectionism and spontaneity, laboring over certain passages and scraping away areas he did not deem

25

successful in order to rework them [4,5]. The extent to which Manet considered this large portrait complete is unknown; however, the predominant aesthetic of the time expected more finished paintings, which might explain the posthumous alterations realized to make this work more marketable.

Figure 1. (**Left**), Édouard Manet, *Woman in Striped Dress*, ca. 1877-80. Oil on canvas, 175.5 × 84.3 cm. Solomon R. Guggenheim Museum, New York, Thannhauser Collection, Gift, Justin K. Thannhauser, 1978, 78.2514.24 © Solomon R. Guggenheim Foundation, New York. Photograph taken in 2015, before conservation treatment. (**Right**), the painting as photographed by Fernand Lochard in the artist's studio (photo no. 118), Paris, ca. 1883-84, 10.5 × 6.5 cm.

A comparison of *Woman in Striped Dress* with a historical image that photographer Fernand Lochard took around 1884 after Manet's death (Figure 1) revealed that the painting suffered a series of dramatic alterations prior to its arrival at the Guggenheim in 1965 as an extended loan, after it was promised as a gift in 1963. Captured at the request of the artist's stepson, Léon Leenhoff, to record the contents of the artist's studio, the Lochard photograph shows, for instance, a rectangular outline drawn around much of the woman's figure. This rough outline suggests that someone—potentially Manet, or else his heirs, friends, or dealers—may have considered reducing the size of the work on all four sides. Accordingly, compared to the original composition as shown in these photographs, the canvas appears to have been cut down to a more vertical format, perhaps in an effort to remove the sketchiest areas of brushwork or to further highlight the model and her finery. Another hand (or hands) also embellished and retouched the work, and numerous additions were made to fill in some incomplete areas of the composition. These changes were likely executed concurrently with the cutting down, as some of the brushwork ends abruptly near the "new" edge of the painting. At a later date, the work lost a few additional centimeters, possibly in conjunction with the attachment of its present lining on the reverse of the canvas, resulting in its current dimensions.

The application of multiple varnish layers, the uppermost of which was immediately evident upon observation of the surface, further compromised the original appearance of *Woman in Striped Dress*. Besides visually flattening the composition, these coatings, which had discolored to a dark greenish-yellow tone over time, partially disguised the retouches. The topmost varnish layer, of a particularly thick and glossy appearance, was likely applied after the Galerie Thannhauser, owned by Justin K. Thannhauser, purchased the painting in 1928. Performed around the same time as the lining treatment, this latest varnishing campaign was perhaps meant to further conceal the discrepancies of the various revisions made to the work. Together with the size reductions and painted embellishments, these coating applications may represent additional attempts to make the picture seem more acceptably finished and therefore desirable to a larger, traditional collector base.

In this context, a comprehensive technical study of Manet's *Woman in Striped Dress* sought to shed light on the original materials the artist used and on the later alterations suffered by the painting during its lifetime, as it passed from one owner to the next. For this purpose, a combination of non-invasive and micro-invasive techniques, as well as portable and benchtop instrumentation, was employed both in the Guggenheim's conservation laboratory and in the Department of Scientific Research (DSR) of The Metropolitan Museum of Art (The Met). In addition to augmenting the current scholarship regarding a pivotal figure of the late 19th century who straddled multiple "isms" while remaining singular in his pictorial production, the findings presented here contributed to the design of a suitable cleaning treatment for the painting, aiming to remove the yellowed superficial varnish and some of the later retouchings. The overall appearance of *Woman in Striped Dress* is now comparable to other Manet portraits of a similar date, with its masterful brushwork and sketchy matte finish. While the work can never be returned entirely to its original scale and condition, it now better captures the spirit of modernity not fully appreciated—certainly in more conservative camps—at the time of its creation.

2. Materials and Methods

The present study of Manet's *Woman in Striped Dress* relied on in situ, non-invasive measurements with portable equipment, followed by the removal and investigation of microscopic samples through benchtop instrumentation. Initial examination of the painting involved an imaging campaign using normal light and infrared (IR) illumination, as well as analysis with X-radiography and handheld X-ray fluorescence (XRF) spectroscopy. After these preliminary investigations, all carried out in the Guggenheim's conservation studio, the work was transferred to The Met for macro-XRF (MA-XRF) analysis. With the invaluable help and guidance of the Guggenheim's conservators, sixteen microscopic scrapings and samples for cross sections were then removed from the woman's dress and from areas of possible later retouching for investigation with optical microscopy and analysis by transmission Fourier-transform infrared (FTIR) and Raman spectroscopies, scanning electron microscopy with energy-dispersive X-ray spectroscopy (SEM/EDS), and electron backscatter diffraction (EBSD). These analyses on the paint stratigraphy were conducted to provide a conclusive identification of the variety of pigments and extenders present in the ground and paint layers. Some of the sample scrapings were also investigated with pyrolysis—gas chromatography/mass spectrometry (Py-GC/MS) for a detailed characterization of the uppermost varnish. Figure 2 reports an indication of the locations where samples were removed for this study, to be referred to from here on whenever scrapings and cross sections are mentioned. Experimental conditions for the analytical techniques employed are reported below.

2.1. IR Reflectography and Transmittography

Imaging was conducted using an Osiris shortwave IR (SWIR) imager (OPUS Instruments, Norwich, UK), equipped with an InGaAs detector with sensitivity in the

900–1700 nm range. An 850 nm long-pass IR filter was used with a Rodagon 150 mm f/5.6 lens, optimized for the IR region. The Osiris SWIR imager has a linear 512-pixel sensor within a precision-geared mechanism that assembles a final 4096 × 4096-pixel image file. Images were post-processed and optimized with Adobe Photoshop software.

2.2. X-Radiography

Imaging was carried out with a Picker Hotshot AXR X-ray system (Associated X-Ray Corporation, New Haven, CT, USA). Images were captured at 45 kV, 3 mA, and 45 s, and were digitized and stitched at the Northeast Document Conservation Center (NEDCC), using a GE model FS50B X-ray scanner. The scanner utilized GE Rhythm Acquire version 4.0 and Rhythm Review version 4.0 to process the files. Stitching and other post-processing work was performed with a combination of PTGui and Adobe Photoshop software.

Figure 2. Sampling sites for Manet's *Woman in Striped Dress*.

2.3. Point XRF

Analysis was performed using a handheld Bruker Tracer III-V^{TM} energy dispersive XRF analyzer (Bruker Corporation, Berlin, Germany), with Peltier-cooled advanced high-resolution silver-free Si-PIN detector with a 0.2 μm beryllium (Be) window and average resolution of approximately 142 eV for the full width at half maximum of the manganese (Mn) Kα line. The system is equipped with changeable filters, and a rhodium (Rh) transmission target with maximum voltage of 45 kV and tunable beam current of 2–25 μA. The size of the spot analyzed is approximately 3 × 4 mm. Experimental parameters were 40 kV, 12.5 μA, and 120 s acquisition time, and a titanium (Ti)–aluminum (Al) filter was used.

Measurements were acquired by positioning the instrument at a $\approx$1 mm distance from the artwork's surface.

2.4. MA-XRF

Mapping was carried out using a Bruker M6 Jetstream® instrument (Bruker Corporation, Berlin, Germany) equipped with a 30 mm^2 XFlash silicon drift detector (SDD) and an air-cooled micro-focus Rh target X-ray tube operated at 50 kV and 0.5 mA. Five areas of the painting were mapped with a 700 μm spot size, an 800 μm step, and a dwell time of 90 ms/pixel. Spectra were processed using the Bruker M6 Jetstream software.

2.5. FTIR

Analysis was conducted in transmission mode with a Hyperion 3000 FTIR spectrometer (Bruker Corporation, Berlin, Germany) equipped with a mercury cadmium telluride (MCT) detector. Each sample was crushed in a Spectra Tech diamond anvil cell and all the materials contained in it were analyzed as a bulk through a 15 × objective. Spectra were collected in the 4000–600 cm^{-1} range, at a resolution of 4 cm^{-1}, as the sum of 128 or 256 scans, depending on the spectral response of the samples examined.

2.6. Raman

Spectra were acquired with a Bruker Senterra Raman spectrometer (Bruker Corporation, Berlin, Germany) equipped with an Olympus 50× long working distance objective and a charge-coupled device (CCD) detector. A Spectra Physics Cyan solid-state laser, a Melles Griot He-Ne laser, and a continuous wave diode laser, emitting light at 488, 633, and 785 nm respectively, were used as the excitation sources, and two holographic gratings (1800 and 1200 rulings/mm) provided a spectral resolution of 3–5 cm^{-1}. The output laser power was kept below 25 mW, while the number of scans and integration time were adjusted to prevent damage from overheating and according to the spectral response of the samples examined.

2.7. SEM/EDS

Analysis of cross sections was performed with a FE-SEM Zeiss Σigma HD system (Zeiss, Oberkochen, Germany) equipped with an Oxford Instrument X-MaxN 80 SDD (Oxford, Tubney Woods, Abingdon, Oxon, UK). Back-scattered electron (BSE) imaging, as well as EDS elemental spot analysis and mapping, were performed in high vacuum with an accelerating voltage of 20 kV, on 12 nm carbon-coated samples.

2.8. EBSD

Examination of a cross section was conducted with the same FE-SEM Zeiss Σigma HD system (Zeiss, Oberkochen, Germany) described above, equipped with an Oxford Instrument X-MaxN 80 SDD and a Nordlys EBSD detector (Oxford, Tubney Woods, Abingdon, Oxon, UK). Patterns were collected at variable pressure (P = 50 Pa), with high current and a 120 μm aperture at 20 kV; the sample was tilted at 70° for analysis. Prior to analysis, the cross section was ion milled under the following conditions: 15 min 2.5 kV accelerating voltage, 1.5 kV discharge voltage (80°); 15 min 1 kV accelerating voltage, 1.5 kV discharge voltage (80°); 15 min 1 kV accelerating voltage, 1.5 kV discharge voltage (70°).

2.9. Py-GC/MS

Measurements were carried out on an Agilent 6890 gas chromatograph (Agilent Technologies, Santa Clara, CA, USA) equipped with a Frontier PY-2020iD Double-Shot vertical furnace pyrolyzer fitted with an AS-1020E Auto-Shot autosampler. The GC was coupled to a 5973N single quadrupole mass selective detector (MSD) (Agilent Technologies, Santa Clara, CA, USA). Samples of 30–50 μg were weighed out in deactivated pyrolysis sample cups (PY1-EC80F Disposable Eco-Cup LF) on a Mettler Toledo UMX2 Ultra microbalance. Samples were then either pyrolyzed without derivatization or derivatized with tetramethyl ammonium hydroxide (TMAH) before pyrolysis. Derivatization took place in the same cups as follows: 3–4 µL of 25% TMAH in methanol (both from Fisher Scientific), depending on the sample size, were added directly to the sample in each cup with a 50 µL syringe and, after 1 min, loaded onto the autosampler. The interface to the GC was held at 320 °C and purged with helium for 30 s before opening the valve to the GC column. The samples were then dropped into the furnace and pyrolyzed at 550 °C for 30 s. The pyrolysis products were transferred directly to a DB-5MS capillary column (30 m × 0.25 mm × 1 µm) with the helium carrier gas set to a constant flow of 1.5 mL/min. Injection with a 30:1 split was used, in accordance with the sample size. The GC oven temperature program was: 40 °C for 1 min; 10 °C/min to 320 °C; isothermal for 1 min. The Agilent 5973N MSD conditions were set as follows: transfer line at 320 °C, MS Quad 150 °C, MS Source 230 °C, electron multiplier at approximately 1770 V; scan range 33–550 amu. For samples run with TMAH, the detector was turned off until 3 min to avoid saturation by excess of the derivatizing agent and solvent. Data analysis was performed on an Agilent MSD ChemStation D.02.00.275 software by comparison with the NIST 2005 spectral libraries.

2.10. Preparation of Cross Sections

Samples were mounted in Technovit 2000 LC resin, cured under ultraviolet (UV) light for 20 min, then polished using Micro-mesh cloths to expose the stratigraphy. Cross sections were examined by means of a Zeiss Axio Imager M2m microscope (Zeiss, Oberkochen, Germany), equipped with an Axiocam HRc digital camera (Zeiss, Oberkochen, Germany) and 50×, 100×, 200×, 400×, and 500× magnifications. Photomicrographs were taken using the AxioVision 4.X.X software.

3. Results and Discussion

The main results of the analysis are summarized below, organized in sections that deal specifically with the canvas support and ground preparation, paint layers, and varnish. Table 1 presents an overview of the data collected from the micro-invasive investigation of the sample scrapings and cross sections removed from the painting.

Table 1. Summary of the compositions obtained with various micro-analytical techniques on sample scrapings and cross sections removed from Manet's *Woman in Striped Dress*. Sample S1 is not reported in the Table as it was too small to be mounted as a cross section.

Samples	Analytical Techniques	Ground Preparation	Pigments, Fillers, and Extenders in Paint Layers	Varnish
(S2) Scraping of uppermost varnish; background, left of center, top quadrant	FTIR, Py-GC/MS	—	—	Diterpenoid natural resin belonging to the Pinaceae family, linseed oil with possible addition of driers
(S3) Cross section of ground; proper right edge, tape removed	Optical microscopy, SEM/EDS	Single-layer ground (20–50 µm): coarse lead white, with a few particles of calcite, barite, feldspar, and carbon-based black	—	One layer (6–20 µm), on top of ground preparation
(S4) Cross section of green paint; foliage, proper right edge, farther into painting	Optical microscopy, SEM/EDS	Single-layer ground (50 µm): coarse lead white, with a few particles of calcite	Top: Calcite, gypsum, Fe oxide/oxy-hydroxide Bottom: Emerald green, ultramarine blue, cadmium yellow, Naples yellow, vermilion, Cr-based green (possibly viridian), lead white, Fe oxide/oxy-hydroxide, organic lake on Al substrate	Two layers (bottom 10–20 µm, top 5–10 µm), on top of paint layers

Table 1. *Cont.*

Samples	Analytical Techniques	Ground Preparation	Pigments, Fillers, and Extenders in Paint Layers	Varnish
(S5) Scraping of green paint; foliage, proper right, top quadrant	FTIR, Raman	—	Viridian, emerald green, lead white (hydrocerussite)	Natural resin and oil (varnish and/or binding medium)
(S6) Scraping of green paint; foliage, right of center, top quadrant	FTIR, Raman	—	Viridian, ultramarine blue, chrome yellow, lead white (hydrocerussite), barite	Natural resin and oil (varnish and/or binding medium)
(S7) Scraping of green paint; foliage, left of center, top quadrant	FTIR, Raman	—	Viridian, ultramarine blue, vermilion, red and yellow ochers, chrome yellow, lead white (cerussite), kaolinite	Natural resin and oil (varnish and/or binding medium)
(S8) Scraping of green paint; foliage, left of center, top quadrant	FTIR, Raman	—	Viridian, ultramarine blue, chrome yellow, lead white (hydrocerussite), barite	Natural resin and oil (varnish and/or binding medium)
(S9) Cross section of brown paint; foliage, proper right edge, tape removed	Optical microscopy, SEM/EDS, EBSD	Single-layer ground (60–70 µm): coarse lead white, with a few particles of calcite and quartz	Top: Lead white, cobalt blue, ultramarine blue, emerald green, malachite, cadmium yellow, Fe oxide/oxy-hydroxide Bottom: Lead white, cadmium yellow, cobalt blue, ultramarine blue, Cr-based green (possibly viridian), emerald green, Fe oxide/oxy-hydroxide, red lake on Al substrate, Naples yellow, malachite, quartz	One layer (3–6 µm), on top of paint layers
(S10) Cross section of brown paint; foliage, proper right edge, farther into painting. Broken into two fragments, 10a and 10b	Optical microscopy, SEM/EDS	Single-layer ground (30 µm): coarse lead white, with a few particles of calcite and gypsum	Top: Calcite, gypsum, Fe oxide/oxy-hydroxide Bottom: Lead white, cadmium yellow, chrome yellow, cobalt blue, ultramarine blue, Cr-based green (possibly viridian), emerald green, Fe oxide/oxy-hydroxide, red lake on Al substrate, vermilion, Naples yellow, Cu-based green (possibly malachite), bone or ivory black, cerulean blue, quartz	Two layers (bottom 6–25 µm, top 3–6 µm), on top of paint layers
(S11) Scraping of green paint; foliage, proper right, top quadrant	FTIR, Raman	—	Viridian, ultramarine blue, vermilion, chrome yellow, lead white (hydrocerussite), barite	Natural resin and oil (varnish and/or binding medium)
(S12) Cross section of pink-orange paint; basket of flowers, near proper right edge	Optical microscopy, SEM/EDS	Single-layer ground (partial): coarse lead white, with a few particles of calcite	Top: Lead white, barite, vermilion, chrome yellow, red lakes on Al and S substrates, emerald green, gypsumBottom: Red ocher, lead white, ultramarine blue, cobalt blue	Three layers: two in between paint layers (bottom 2–5 µm, top 4–5 µm); one at top of stratigraphy (8–10 µm)
(S13) Cross section of green paint; foliage, proper right, top quadrant	Optical microscopy, SEM/EDS	Single-layer ground (10–40 µm): coarse lead white, with a few particles of calcite, barite, silicates, and iron-containing earths	Top: Cr-based green (possibly viridian), ultramarine blue, chrome yellow, lead white, barite, Fe oxide/oxy-hydroxide, vermilion, calcite Bottom: Emerald green, lead white, Fe oxide/oxy-hydroxide, ultramarine blue, vermilion, Naples yellow, zinc yellow, bone or ivory black, calcite, barite	Two layers: one in between paint layers (5–15 µm); one on top of paint layers (20–40 µm)

Table 1. *Cont.*

Samples	Analytical Techniques	Ground Preparation	Pigments, Fillers, and Extenders in Paint Layers	Varnish
(S14) Cross section of green paint; foliage, proper left, top quadrant	Optical microscopy, SEM/EDS	—	Cr-based green (possibly viridian), chrome yellow, lead white, barite, red ocher, calcite	None observed
(S15) Scraping of blue paint; dress, near proper left edge, bottom quadrant	Raman	—	Ultramarine blue, carbon-based black	—
(S16) Scraping of blue paint; dress, center of picture	Raman	—	Ultramarine blue	—

3.1. Canvas Support and Ground Preparation

A comparison between the Lochard photograph and *Woman in Striped Dress* as seen in the Guggenheim's galleries (Figure 1) reveals that the picture underwent noticeable alterations before it arrived at the museum in 1965, as mentioned above. While retained by Manet's widow, Suzanne Manet, between 1883 and 1902, the painting was cut down on the left- and right-hand sides and trimmed at the top. The original size of the work was approximately 192 × 124 cm; by 1902 it measured about 180 × 85 cm; the current dimensions are 175.5 × 84.3 cm [1]. While in the present case the change in canvas size appears to have been carried out after the artist's death, there are instances in which Manet himself would willfully extend his pictures by adding a piece of canvas to one side, divide a work into separate compositions and subsequently rejoin selected parts, or trim a painting from the top and bottom to adjust its shape [6]. *Woman in Striped Dress* is double lined with an aqueous adhesive and mounted onto a non-original, rather heavyweight stretcher, deliberately stained to a reddish-brown color. Traces of old tack holes along the edges, as well as microscopic cracking lines approximately 5 mm in from the existing cut edge, indicate that the canvas might have previously been folded around a slightly smaller stretcher than the one on which it is mounted today. At some point, tape was applied all around the edges of the picture—a practice that was often considered a neat way of completing a lining and is still in use today. Upon visual examination, the painting displays three small tears that may have prompted its lining, respectively located in the lower left of the floor area, in the skirt to the left of the model's proper left hand, and in the upper right corner.

Manet executed the painting on a canvas with a particularly fine weave considering its large size. Evidence of a preparatory drawing was not detected in the IR reflectogram, and the X-radiograph indicated that most of the composition was unchanged (Figure 3). Both suggest a remarkably direct and confident execution. The thin, off-white ground allows the texture of the canvas weave to remain prominent, a practice often favored by the Impressionists, with whom Manet was associated. SEM/EDS analysis of cross sections S3, S4, S9, S10, S10b, and S12 showed that the ground, varying in thickness between 20 and 70 μm, consists mainly of coarse lead white, with a few particles of calcite, gypsum, barite, quartz, feldspar, and a carbon-based black (Figure 4).

Figure 3. From left to right, IR reflectogram and X-radiograph of Manet's *Woman in Striped Dress*.

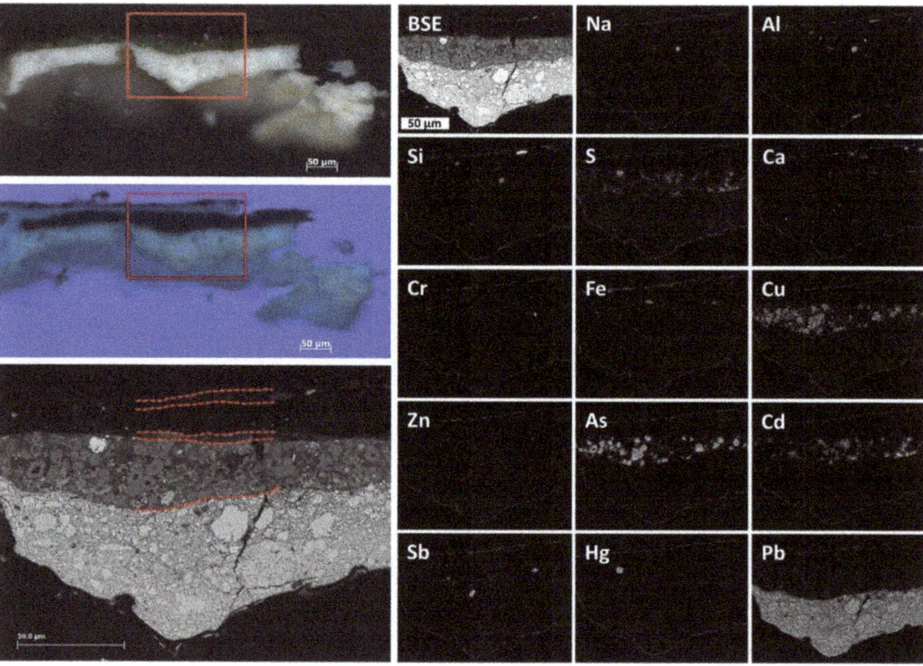

Figure 4. (**Left**) Photomicrographs of cross section S4 taken under polarized visible and UV light illuminations, respectively, and BSE image. S4 was removed from an area of foliage in the painting's proper right edge. In the photomicrographs, the red rectangles indicate the area where the BSE image and EDS elemental maps were acquired. (**Right**) EDS elemental maps of sodium (Na), aluminum (Al), silicon (Si), sulfur (S), calcium (Ca), chromium (Cr), iron (Fe), copper (Cu), zinc (Zn), arsenic (As), cadmium (Cd), antimony (Sb), mercury (Hg), and lead (Pb).

3.2. Paint Layers

As indicated above, a comparison of the portrait and Lochard photograph (Figure 1) betrayed painterly embellishments added by another hand or hands, especially over areas of exposed ground in the background, the model's right hand, and the decoration on the tabletop. In particular, while Manet originally only hinted at the background, the retouches added more detailed foliage, flowers, and a distinct rigid trellis-like pattern. These additions stood in striking contrast to Manet's deft brushwork and compressed the space around the main figure, resulting in a flattening effect. Moreover, the Louis XV side table, recurring in many of the artist's works, was originally located at a diagonal; the canvas reduction, which removed the table legs on the left, required that the table be seen almost straight on. As a result, the decorative element at the center of the table became one at its corner, and the suggestion of a leg was added, shifting the whole angle of the table within the composition. Besides altering the perspective, removal of the table legs from the left of the picture caused the woman, who was originally perched on or leaning against the table, to appear to be standing in a more precarious position. The graceful gesture of her proper right, gloved hand was also changed, resulting in a stiff, pointed extremity. Likewise, the lower right section of the model's skirt, which was only summarily sketched by Manet, was obscured at a later date with a thin, warm black paint, probably in an effort to cover the exposed ground and make this passage appear more finished. The model's proper right eyebrow, moreover, initially raised in a potentially mischievous fashion, was replaced with a thin mark that softened her visage and suppressed any alluring expression.

An inconspicuous calligraphic inscription that reads "Ed Manet" is evident on the table at center left. This inscription differs from the more usual robust signature found on other paintings by the artist, including *Before the Mirror* (1876), also at the Guggenheim (Suppl. Figure S1) [7]. When and by whom this inscription was applied is not known, although it appears to be directly on top of the original paint.

Thinner and washier in the background, the paint is characterized by some modest impasto in locations corresponding to the focus of the composition—the female figure and her fashionable garment, her fan, and the decorative carving on the table. Combined analysis of sample scrapings and cross sections by SEM/EDS, Raman, and FTIR spectroscopies showed that the various fields of color typically consist of numerous pigments, many of which are detected in trace amounts and may be unintentional, picked up from an unwashed brush or a much-used palette. The finding of complex mixtures, often including up to ten individual pigments within a single paint layer, is in keeping with previous studies of this artist's work by other research groups [4]. Materials identified in the ground and paint layers are consistent with those found in paintings by Manet in other collections [4,6,8,9].

The foliage in the painting's background was identified as a critical area of possible posthumous reworking, which required in-depth analysis. Five microscopic scrapings (S5, S6, S7, S8, S11) were removed from different green shades in the top quadrant. Among these, sample S5 was found to be a mixture of viridian and emerald green; S6 and S8 contain viridian, ultramarine blue, and a lead chromate-based pigment (Figure 5); S7, of a distinctive olive green tone, is composed of the same three pigments, with the addition of vermilion, as well as red and yellow earths; S11 is based on viridian, ultramarine blue, a lead chromate-based pigment, and vermilion. All of these paint mixtures contain lead white, mostly present in the form of hydrocerussite; in addition, some include barite (S6, S8, S11) and kaolinite (S7) (Figure 5, right), probably as extenders. FTIR spectra of samples S6, S7, S8, and S11 also display a series of bands that are typical of metal carboxylates (Figure 5, right), most likely arising from the chemical interaction between the paint and oil medium.

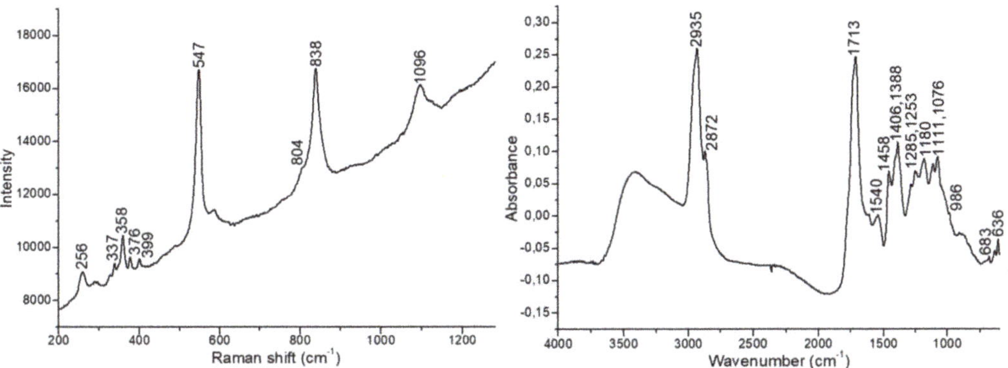

Figure 5. (**Left**) Raman spectrum of sample S6, i.e., a scraping of light green paint from an area of foliage in the painting's background at proper right, displaying the characteristic bands of ultramarine blue and a lead-chromate pigment. (**Right**) FTIR spectrum of the same sample, showing the distinctive signals of viridian, lead white in the form of hydrocerussite, barite, and metal carboxylates.

In addition to scrapings, seven cross sections (S4, S9, S10, S10b, S13, S14b, S14c) were taken from green and brown passages of foliage to study the paint stratigraphy. Microscopic observation of the cross sections under visible polarized light illumination revealed the structure of Manet's paint to be, mostly, a single layer applied directly over the ground, which the artist left visible in many passages. In sample S4, removed from an area of foliage in the painting's proper right edge, the green paint consists predominantly of emerald green, mixed with ultramarine blue, cadmium yellow, Naples yellow, a few particles of vermilion, a chromium-based green, possibly viridian, and traces of lead white, an iron-based compound, most likely an oxide or oxy-hydroxide, and aluminum-containing particles that are possibly the substrate of an organic lake (Figure 4). In cross section S9, taken from an area of the basket of flowers on the proper right edge where the tape had been removed, the green paint is a mixture of lead white, cadmium yellow, cobalt blue, ultramarine blue, a chromium-based green, possibly viridian, emerald green, an iron-based compound, a red lake precipitated onto an aluminum-based substrate, Naples yellow, a copper-based green, and a few quartz grains (Supplementary Figure S2). This copper-based green was confirmed as malachite by means of EBSD analysis, consistently with its particle morphology and chemical composition (Supplementary Figure S3). The same green layer detected in S9 is also present in S10 and S10b, collected from a location close to S9 yet farther into the painting. While the overall composition is the same, additional compounds were identified in these two specimens—namely, chrome yellow, vermilion, cerulean blue, and a bone or ivory black. In S9, the green layer was modified by the addition, most likely wet-on-wet, of a paint layer rich in fine particles of lead white, as well as a few particles of the same pigments found in the underlying layer; among them, cobalt blue, ultramarine blue, emerald green, malachite, cadmium yellow, and an iron-based compound, most likely an oxide or oxy-hydroxide. In S4, S10, and S10b, a discontinuous, relatively thin layer (3 to 10 μm) containing small, discrete particles of calcite and gypsum, as well as a few iron-containing particles, is present over the green paint.

Cross sections S4, S9, S10, and S10b, discussed above, are composed of a single layer of green paint, which is always located on top of the ground preparation and covered by a single or double varnish application (shown in Figure 4 for S4 and in Supplementary Figure S2 for S9). Conversely, sample S13, removed from an area of foliage in the painting's background near the proper right edge, appears to contain two distinct green paint layers: one applied directly over the ground; and a second one in the upper portion of the stratigraphy, sandwiched between two varnish applications (Figure 6). SEM/EDS analysis showed that the lowermost of these green paint layers is composed of a mixture of emerald green, lead white, iron oxide/oxy-hydroxides, ultramarine blue, as well as traces of vermilion, Naples yellow, zinc yellow, a bone or ivory black, calcite, and barite. The uppermost green layer contains a chromium-based green, possibly viridian, ultramarine blue, chrome yellow, and lead white, with abundant barite, as well as iron oxide/oxy-hydroxides, a few minuscule vermilion and calcite particles (Figure 6). The top paint layer was also observed and analyzed in cross sections S14b and S14c, taken from an area of green foliage at the opposite side of the painting, near its proper left edge. A combined elemental map showing the distributions of chromium, mainly concentrated in the upper paint layers as a chrome yellow and a chromium-based green, possibly viridian, and of copper, due to the emerald green present in the bottom paint layers, is presented in Figure 7.

While paint layers applied directly over the ground are presumed to be original, those found on top of a varnish, as in the case of S13, likely point to later additions. All pigments detected in the sample scrapings and cross sections of green paint were known and available at the end of the 19th century [10], indicating that, albeit posthumous with respect to the original composition, the uppermost paint in S13 could have been applied shortly after the artist's death.

Comparison with the Lochard photograph showed that the basket of flowers, too, was likely reworked after Manet's death. As such, this site was deemed of particular interest and deserving investigation. Photomicrographs taken under visible and UV illuminations of cross section S12, removed from one of the pink-orange petals, revealed a complex stratigraphy, including, above the ground, a yellow-brown layer, followed by two varnish applications, a pink-orange layer, and one additional varnish application (Supplementary Figure S5). As mentioned above, the observed location of the yellow-brown paint, applied directly over the ground, points to the possibility of this being original paint; conversely, the fact that the pink-orange paint is lying on top of two varnish layers indicates a possible later addition. The yellow-brown paint is composed of an iron-based compound and silicates, most likely present as a red ocher, along with lead white, ultramarine blue, and a few cobalt blue particles. The pink-orange paint contains lead white, barite, vermilion, chrome yellow, two red lakes—one with aluminum and sulfur and, possibly, one with a tin-based substrate, and a few emerald green and gypsum particles (Supplementary Figure S5). As observed in the samples of green paint, all pigments detected in S12 were available at the time in which the work was created [10]; thus, these later embellishments could have been introduced soon after Manet's death.

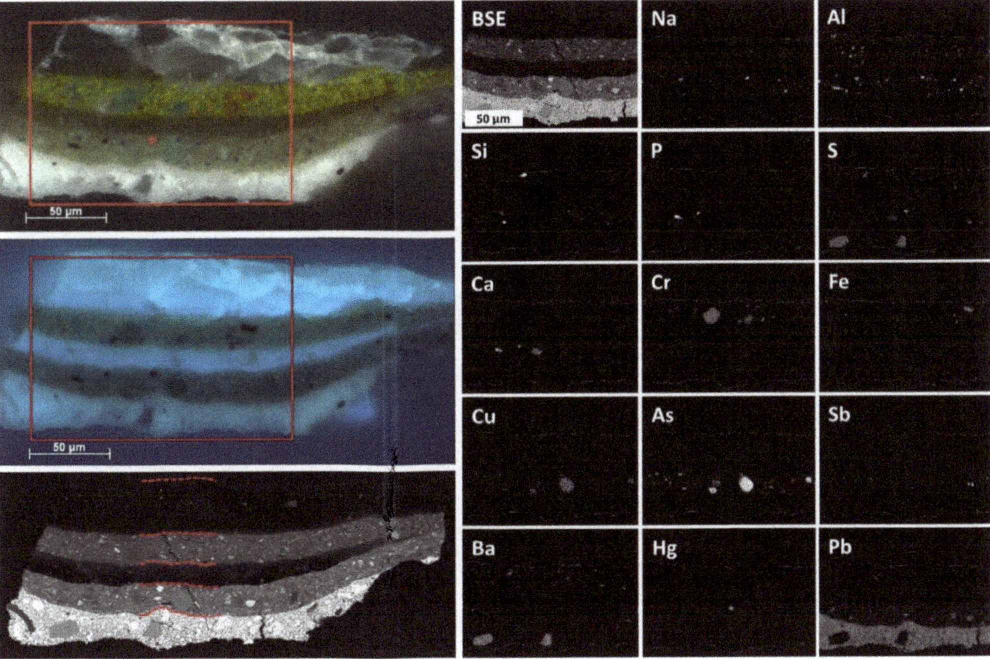

Figure 6. (Left) Photomicrographs of cross section S13 taken under polarized visible and UV light illuminations, respectively, and BSE image. S13 was removed from an area of foliage in the painting's background near the proper right edge. In the photomicrographs, the red rectangles indicate the area where the BSE image and EDS elemental maps were acquired. **(Right)** EDS elemental maps of sodium (Na), aluminum (Al), silicon (Si), phosphorus (P), sulfur (S), calcium (Ca), chromium (Cr), iron (Fe), copper (Cu), arsenic (As), antimony (Sb), barium (Ba), mercury (Hg), and lead (Pb).

Figure 7. Combined elemental distribution maps acquired by MA-XRF in the topmost quadrant of the painting: copper (Cu, light blue) and chromium (Cr, yellow). The area where these maps were acquired is indicated by a red rectangle in Supplementary Figure S4, in the Supplementary Information file.

MA-XRF elemental maps revealed that the grayish-white and black-looking stripes of the dress were painted with lead white and a calcium-containing black pigment (Figure 8, left). Moreover, chromium and copper are also present in these areas, suggesting a more complex pigment mixture. Raman analysis of two samples removed from distinct areas of the dress, S15 and S16, showed that its dark coloration is mainly due to ultramarine blue, and that Manet combined this pigment with a bone black or ivory black for the darker blue strokes outlining the skirt in the lower right. The woman's face was created with just a few assured brushstrokes over a thin, earth-colored, dry scumble of vermilion and iron-containing earths (Figure 8, right). The MA-XRF map shown in Figure 8 at right also revealed that her eyes were painted primarily using cobalt blue, with a hint of brown paint—again, likely an iron-containing earth—in her proper left eye and that her lips were made with vermilion, although traces of several other pigments were added to subtly adjust the color shades.

Figure 8. Combined elemental distribution maps acquired by MA-XRF in two areas of the painting. (**Left**) Area of the sitter's dress: lead (Pb, white), chromium (Cr, yellow), copper (Cu, green), and calcium (Ca, violet); (**Right**) Area in the upper right quadrant: cobalt (Co, blue), iron (Fe, orange), and mercury (Hg, red). The areas where these maps were acquired are indicated by green and blue rectangles in Supplementary Figure S4, in the Supplementary Information file.

3.3. Varnish

Examination with optical microscopy and SEM/EDS analysis of sample cross sections provided useful information on the number of varnish layers present in various areas of the picture and their locations within the stratigraphy. Sample S3, for instance, was taken from a site on the painting's proper right edge where the tape had been removed. Photomicrographs of this cross section display a 6 to 20 μm thick varnish on top of the ground preparation (Supplementary Figure S6), which is thought to be an initial coating applied before any of the embellishments were added to the composition and tape was applied around the work's edges. Sample S4, collected from a location on the proper right edge, close to S3, although farther into the painting, shows two varnish layers, the uppermost of which has an average thickness of 5–10 μm and was likely applied after the tape and within its borders. Similar observations to those made for samples S3 and S4 apply to S9 and S10, taken from a lower portion of the painting's proper right edge—the first from a site where the tape had been removed, the latter (broken into two fragments, S10 and S10b) farther into the composition. In these cases, too, the samples display one and two varnish layers, respectively, the uppermost one appearing generally thinner than the lowermost application. Flakes of a copper and zinc alloy detected in samples S10 and S10b are interpreted as residues from the brass gilding of the painting's historic frame.

Transmission FTIR analysis of a scraping of the most superficial varnish removed from an area in the painting's background at top (S2) detected mostly a natural resin (Figure 9, left). Py-GC/MS analysis provided a more detailed characterization of the sample components, showing the presence of several oxidation products of abietic acid, including methyl abietate, methyl 6-dehydrodehydroabietate, 7-methoxytetrahydroabietic acid, 15-methoxydehydroabietic acid, 7-oxodehydroabietic acid, 7,15-dimethoxytetradehydroabietic acid, and 15-methoxydehydroabietic acid (Figure 9, right). These compounds are indicative of a diterpenoid natural resin belonging to the Pinaceae family [11]. In addition, glycerol along with azelaic, palmitic, and stearic acids were identified, suggesting the presence of an oil [12]. While fatty acid ratios cannot always be relied upon to provide a detailed characterization of the oil, in this case, the relative ratio of palmitic acid to stearic acid (P/S = 1.65) suggests the presence of linseed oil, while the slightly elevated relative ratio of azelaic acid to palmitic acid (A/P = 1.9) may indicate the addition of driers. More frequently employed for wood furniture or musical instruments rather than modern paintings, a varnish of such composition has been often described as an inexpensive alternative to more traditional artists' varnishes [13].

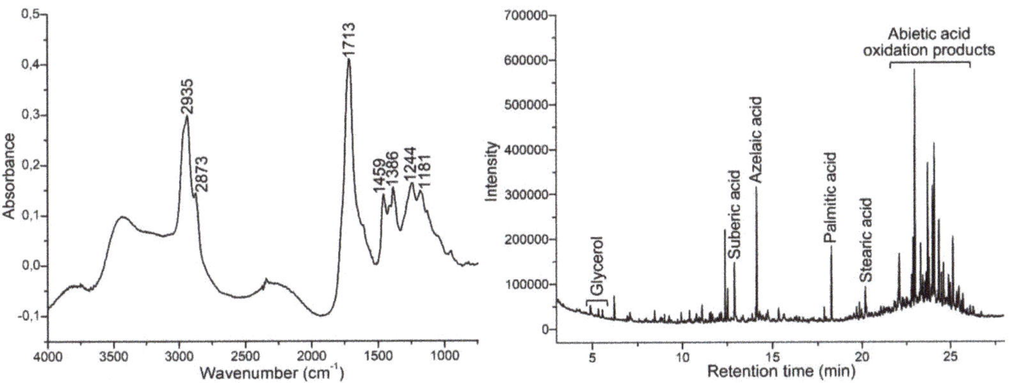

Figure 9. (**Left**) FTIR spectrum of sample S2, i.e., a scraping of the uppermost varnish from an area in the painting's background at top, displaying the characteristic bands of a natural resin. (**Right**) Py-GC/MS chromatogram of sample S2 obtained with TMAH derivatization. Materials identified include Pinaceae resin and oil. Marker compounds were detected as methylated derivatives.

4. Conclusions

This article describes the salient results of a technical study on *Woman in Striped Dress*, a large-scale portrait by Édouard Manet in the Thannhauser collection of the Solomon R. Guggenheim Museum, New York. The goal of this research, a collaboration between the Guggenheim's Conservation and Curatorial Departments and The Met's DSR, was to illuminate the original materials and posthumous alterations of this masterpiece in support of an ongoing conservation treatment. Scientific analysis, along with comparative examination of today's picture with a photograph taken by Fernand Lochard after Manet's death, uncovered substantial changes that had been made to the painting before it was promised as a gift to the Guggenheim in 1963, first arriving at the institution as an extended loan in 1965. Among these are subsequent reductions in size, several retouches and embellishments likely motivated by a desire to make the work appear more "finished", thus marketable, as well as at least three varnishing campaigns. Overall, these revisions had severely compromised the appearance of Manet's picture.

The examination of cross section samples removed from locations possibly affected by later retouches revealed the presence of at least one intermediate varnish layer, confirming the posthumous nature of the superficial paint application. However, all materials in the ground and paint layers are consistent with those found in paintings by the same artist in other collections and were in use in the second half of the 19th century, which indicates that these later embellishments may have been introduced shortly after Manet's death. Analysis of a sample scraping of the uppermost varnish layer detected a natural diterpenoid resin of the Pinaceae family mixed with linseed oil—a more common formulation for wooden objects rather than modern paintings. In addition to analysis and observation with UV light and stereomicroscopy, the solubility parameters of the varnishes, retouching, and original paint layers were quite distinct and aided in the identification of dissimilar materials. This combined information was crucial to devise and guide a conservation treatment for the painting, which took place between 2015 and 2018 in an attempt to remove the discolored varnish and select areas of retouching. This treatment approach privileged the artist's hand, while retaining and integrating some historical—albeit later—elements in a holistic way to preserve the painting's overall aesthetic (Figure 10).

Figure 10. Three areas of the painting, from left to right details of the corsage, dress and glove, shown before (**top**) and after (**bottom**) cleaning. © Solomon R. Guggenheim Foundation, New York.

Upon varnish removal, some of the paint strokes in areas of later retouching appeared dense and flat, in contrast with Manet's inspired technique, always characterized by bright and broad brushwork, even in the presence of complex pigment mixtures. After cleaning, the object on the table, historically cast as a basket of flowers, was reinterpreted as, possibly, a bouquet or a hat with silk flowers. Notably, the woman's dress, which had for decades read as black and off-white, was, after treatment, revealed to be of a nuanced palette of grayish-white and black with a deep blue-violet hue. These colors concur with Théodore Duret's description in his 1902 monograph on Manet, which characterized the dress as striped *"grises-violettes"* [14]. While the painting may never be restored to its original scale and state, the scientific work and conservation treatment described in this article contributed to the return of *Woman in Striped Dress* to a closer approximation of its original appearance.

Supplementary Materials: The following are available online at https://www.mdpi.com/article/10.3390/coatings11111334/s1, Figure S1: Inscription found on Woman in Striped Dress (left) compared to Manet's signature on Before the Mirror (right), Figure S2: Optical microscopy and SEM/EDS data acquired on cross Section S9, Figure S3: EBSD data acquired on cross Section S9, Figure S4: Areas of the painting where the MA-XRF elemental distribution maps shown in Figures 7 and 8 were acquired, Figure S5: Optical microscopy and SEM/EDS data acquired on cross Section S12, Figure S6: Optical microscopy and SEM/EDS data acquired on cross Section S3.

Author Contributions: F.P. coordinated the study, removed samples, prepared cross sections, carried out optical microscopy, point XRF, FTIR, Raman, Py-GC/MS, as well as part of SEM/EDS analysis and data interpretation, and wrote the initial draft of this manuscript with inputs from all other authors. S.A.C. acquired, processed, and interpreted the MA-XRF data. F.C. conducted part of the SEM/EDS and EBSD analysis and data interpretation. S.A.C. and F.C. contributed to the overall interpretation of the micro-analytical results obtained on the paint samples. G.M. V.G., and L.S. provided art historical context, treatment information, data from IR reflectography, X-radiography, microscopic examination of the painting, and supported the scientific work. This article includes excerpts from a conservation entry and an art historical essay by G.M., V.G. and F.P., and significantly expands the materials information gleaned from scientific analysis. All authors have read and agreed to the published version of the manuscript.

Funding: This research was made possible by the Network Initiative for Conservation Science (NICS), a Metropolitan Museum of Art program. Support for NICS was provided by a grant (31500630) from The Andrew W. Mellon Foundation. Conservation on this painting was made possible by a grant to the Solomon R. Guggenheim Museum from the Bank of America Conservation Project.

Institutional Review Board Statement: Not applicable.

Informed Consent Statement: Not applicable.

Data Availability Statement: All data generated during this study are either included in this published article or available from the corresponding author upon reasonable request.

Acknowledgments: The authors are indebted to Charlotte Hale and Michael Gallagher in The Met's Department of Paintings Conservation for generously supporting this project. The authors are also grateful to Brunella Santarelli and Louisa Smieska, former fellows in The Met's DSR, for assisting with the EBSD and MA-XRF analysis, respectively. In addition, FP would like to acknowledge Anna Cesaratto for her initial contribution to this project during her time at The Met working in the NICS program. The Guggenheim authors wish to thank, most especially, Juliet Wilson-Bareau, Anne Distel, and Charles F. Stuckey. Among our current and former Guggenheim colleagues, we are indebted to Julie Barten, Tracey Bashkoff, Susan Davidson, and Megan Fontanella. We owe special gratitude to David and Zahira Bomford, Aviva Burnstock, and Maureen Cross. We also thank Scott Allen, Carol Armstrong, Elsa Badie Modiri, Vivian Endicott Barnett, Barbara Buckley, Helen Burnham, Nick Brandreth, Jacklyn Chi, S. Hollis Clayson, Nicholas Eastaugh, Thierry Ford, Isabel Gaëtan, Gloria Groom, Sven Haase, Jean-Gabriel Lopez, Paul Messier, Dianne Dwyer Modestini, Doug Munson, Debra Hess Norris, Mark Osterman, Isolde Pludermacher, Aileen Ribeiro, Marie Robert, Betsy Rosasco, Valerie Steele, Susan Alyson Stein, MaryAnne Stevens, Françoise Tétart- Vittu, Jennifer Thompson, John Vincler, Christina Zuccari, and Frank Zuccari.

Conflicts of Interest: The authors declare no conflict of interest.

References

1. McMillan, G.; Pozzi, F. Édouard Manet, Woman in Striped Dress, Materials and Process. In *Thannhauser Collection: French Modernism at the Guggenheim*; Fontanella, M., Ed.; Guggenheim Museum Publications: New York, NY, USA, 2018; pp. 108–110, 298.
2. Greene, V.; McMillan, G. Revealing Édouard Manet's Woman in Striped Dress. In *Thannhauser Collection: French Modernism at the Guggenheim*; Fontanella, M., Ed.; Guggenheim Museum Publications: New York, NY, USA, 2018; pp. 111–115, 298–300.
3. Proust, A. Édouard Manet (Souvenirs). *La Rev. Blanche* **1897**, *12*, 132–133.
4. Hanson, A.C. *Manet and the Modern Tradition*; Yale University Press: New Haven, CT, USA, 1977; p. 160.
5. Stuckey, C.F. Manet Revised: Whodunit? *Art Am.* **1983**, *71*, 158–177, 239–241.
6. Bomford, D.; Roy, A. Manet's "The Waitress": An Investigation into its Origin and Development. *Natl. Gallery Tech. Bull.* **1983**, *7*, 3–19.

7. McMillan, G. Édouard Manet, Before the Mirror, Materials and Process. In *Thannhauser Collection: French Modernism at the Guggenheim*; Fontanella, M., Ed.; Guggenheim Museum Publications: New York, NY, USA, 2018; pp. 104–106.
8. Groom, G.; Westerby, G. (Eds.) Manet Paintings and Works on Paper at the Art Institute of Chicago. 2017. Available online: https://publications.artic.edu/manet/reader/manetart/ (accessed on 5 August 2021).
9. Amato, S.R.; Burnstock, A.; Cross, M.; Janssens, K.; Rosi, F.; Cartechini, L.; Fontana, R.; Fovo, A.D.; Paolantoni, M.; Grazia, C.; et al. Interpreting Technical Evidence from Spectral Imaging of Paintings by Édouard Manet in the Courtauld Gallery. *X-Ray Spectrom.* **2019**, *48*, 282–292. [CrossRef]
10. Eastaugh, N.; Walsh, V.; Chaplin, T.; Siddall, R. *Pigment Compendium. A Dictionary and Optical Microscopy of Historic Pigments*; Butterworth-Heinemann: Oxford, UK, 2004.
11. van den Berg, K.J.; Boon, J.J.; Pastorova, I.; Spetter, L.F. Mass Spectrometric Methodology for the Analysis of Highly Oxidized Diterpenoid Acids in Old Master Paintings. *J. Mass Spectrom.* **2000**, *35*, 512–533. [CrossRef]
12. Mills, J.; White, R. *Organic Chemistry of Museum Objects*, 2nd ed.; Butterworth-Heinemann: Oxford, UK, 1994.
13. White, R.; Kirby, J. A Survey of Nineteenth- and Early Twentieth-Century Varnish Compositions Found on a Selection of Paintings in the National Gallery Collection. *Natl. Gallery Tech. Bull.* **2001**, *22*, 64–84.
14. Duret, T. *Histoire d'Édouard Manet et de Son Oeuvre*; H. Floury: Paris, France, 1902.

Article

Er:YAG Laser Cleaning of Painted Surfaces: Functional Considerations to Improve Efficacy and Reduce Side Effects

Cong Wang [1,2], Yijian Cao [3,*], Fude Tie [1] and Mara Camaiti [4,*]

[1] Key Laboratory of Cultural Heritage Research and Conservation, Northwest University, Ministry of Education, Xi'an 710127, China; congwang@nwu.edu.cn (C.W.); fude_tie@hotmail.com (F.T.)
[2] School of Cultural Heritage, Northwest University, Xi'an 710127, China
[3] Institute of Culture and Heritage, Northwestern Polytechnical University, Xi'an 710072, China
[4] CNR-Institute for Geosciences and Earth Resources, 50121 Florence, Italy
[*] Correspondence: yijian.cao@nwpu.edu.cn (Y.C.); mara.camaiti@igg.cnr.it (M.C.)

Abstract: The restoration of paintings always involves the removal of darkened superficial layers, which are mainly due to dust deposition and aged varnishes. As cleaning is an irreversible and invasive treatment, physical methods (i.e., laser cleaning) instead of chemical ones are frequently suggested to reduce side effects on pictorial layers. Among the most employed laser systems, the free-running Er:YAG laser is considered very suitable for fine arts cleaning. This laser works at 2.94 μm, at which only –OH and –NH bonds in molecules are excited. This character can become a disadvantage when pigments with these functional groups are present. To understand the potential of the Er:YAG laser in such situations or in the presence of degradable pigments, the effectiveness of varnish removal from paintings prepared with egg yolk as the binder and cinnabar and lead white as the pigments were systematically investigated. Different cleaning conditions were used, and a hyperspectral sensor was innovatively used as a rapid, in situ and non-destructive technique to assess the effects of laser ablation, besides microscopic analysis. Though results obtained show all these pigments are sensitive to this laser radiation, satisfactory cleaning can be achieved without damaging the pictorial layer. The best cleaning conditions were 0.5 W of power (50 mJ and 10 Hz for energy and frequency), with 2-propanol as the wetting agent.

Keywords: laser cleaning; Er:YAG laser; paintings; efficacy evaluation

Citation: Wang, C.; Cao, Y.; Tie, F.; Camaiti, M. Er:YAG Laser Cleaning of Painted Surfaces: Functional Considerations to Improve Efficacy and Reduce Side Effects. *Coatings* **2021**, *11*, 1315. https://doi.org/10.3390/coatings11111315

Academic Editors: Ivan Jerman and Maurizio Licchelli

Received: 31 August 2021
Accepted: 26 October 2021
Published: 28 October 2021

1. Introduction

In order to enhance the color saturation as well as to protect the original pictorial layers, a thin and transparent film (typically less than 100 μm) of organic substances is applied as the finishing layer of paintings (mainly easel paintings and also sometimes mural paintings) [1]. In historical and modern times, both natural substances (e.g., dammar, mastic, shellac resins) and synthetic polymers (e.g., P B67, ketone resins) are used as varnish [1,2]. However, owing to their organic nature, varnishes are vulnerable to natural aging triggered by temperature, relative humidity fluctuations and UV irradiation, which results in darkened, rigid and brittle superficial films. In some cases, micro-cracks of varnish are present. Besides, dust, soot and dirt deposition on varnish also contribute negatively to the precise appreciation of the aesthetic features of artworks. As a consequence, when it comes to the restoration of paintings, the removal or cleaning of darkened superficial layers, either due to dust, dirt deposition or aging of varnish, has always been a major focus for restorers.

Since varnish removal is invasive and irreversible, particular attention must be paid, and all methods shall be properly evaluated before application. In tradition, mechanical and chemical methods are usually adopted. Mechanical cleaning is performed by using a scalpel, and the effect of cleaning mainly depends on the skills of the operator. As the boundary between optimum cleaning and over-cleaning is blurring, damage to paint

43

layers is of great possibility. In fact, chemical methods (or wet methods) that involve the use of organic solvents, e.g., acetone, ethanol, toluene, hexane, etc., were preferred by painting restorers. By simply wetting, dissolving and wiping the targeted material with a cotton swap, chemical methods are a rather convenient way for varnish removal. However, adverse effects of chemical cleaning are also conspicuous. For oil paintings, organic solvents can induce a series of physicochemical alterations called "solvent action", including swelling of paint films, diffusion and evaporation of solvents, and leaching of pigments and binders off paint films [3]. Such alterations can be temporary (e.g., swelling effect, metal soap formation) and/or permanent (e.g., redistribution/extraction of components and the leaching of components increased brittleness, while the decreased thickness of paint films changed optical properties of overall painting) [3–6]. Last but not least, the volatility and toxicity of organic solvents should not be neglected, which are harmful to operators and the environment.

Thanks to the rapid advancements of science and technology, novel alternative cleaning methods have been developed in recent decades. In general, there are chemical cleaning methods by physical/chemical gels and physical cleaning methods by laser systems [7–10]; in this study, we focus on the latter. After the invention of the laser device in the 1960s, the first application of lasers to artworks was in 1971 when John Asmus and his team used a 690 nm ruby laser to acquire holographic images of marble statues in Venice [11]. In the experiment, he not only demonstrated the feasibility of holography but also illustrated the possibility of removing encrustation on stones by a ruby laser. His pioneering work with laser started a new field of research in conservation science. Ever since, there have been mainly three types of lasers employed in art conservation, namely the excimer laser, Nd:YAG laser and Er:YAG laser. In comparison with chemical and mechanical methods, laser cleaning shows various advantages that have been summarized as accuracy, selectivity, safety, control and reliability [10].

With variable emission wavelength (UV, NIR), pulse duration and relatively high energy, excimer lasers and Nd:YAG lasers are more suitable for cleaning rigid, inorganic substances such as stones, metals, glasses and wall paintings [11–19]. Due to their high energy (>1 eV) and deep penetration capability, excimer laser and Nd:YAG laser can introduce thermal and mechanical damages to adjacent material surrounding the impact zone, via the so-called "physical amplification" and "delayed amplification" [20]. These characteristics impede their application on delicate, sensitive artifacts. Fortunately, the more recent Er:YAG laser works at 2.94 μm, at which only –OH and –NH bonds in water and other molecules can absorb [21]. Thus, the pattern of energy deposition (actual ablation area) is determined by the distribution of molecules containing –OH and –NH bonds, and consequently, the laser beam is quickly absorbed, achieving very superficial ablation depths. In theory, the efficiency of laser ablation is proportional to the amount of –OH and –NH groups present in the materials. These –OH and –NH bonds, when not present in the original material, can be obtained by the addition of hydroxylated liquids (wetting agents), which can also help to limit the radiation and heat penetration [9,21,22]. These features make the Er:YAG laser particularly suitable for materials with high sensitivity and complex structures, e.g., easel paintings.

Although the Er:YAG laser was successfully applied for varnish removal [23–29], to our best knowledge, there is limited study on the situation when pigments in pictorial layers are also sensitive to this laser. Besides, laser cleaning is thought to be intuitive, and controlled cleaning can be performed by simply changing the fluence or pulse duration of the laser beam based on the experience or skill of the operator. Therefore, finding a more scientific and in situ method to indicate the extent of the cleaning process is imperative and urgent.

In this study, a free-running Er:YAG laser with different working parameters and conditions, e.g., power, energy, frequency, wetting agents, etc., were employed to clean varnishes on painting mock-ups prepared with laser-sensitive pigments (cinnabar, lead white). During the cleaning, besides the traditional optical microscopy, a portable and non-invasive hyperspectral sensor (ASD-FieldSpec FR Pro, NY, USA) that can characterize organic and inorganic materials [30–33] was innovatively exploited to characterize painting components and to monitor the progress of varnish cleaning by the Er:YAG laser.

2. Materials and Methods

2.1. Painting Mock-Ups

Painting mock-ups were prepared on the canvas (Zecchi, Florence, Italy), which was already treated with gypsum and animal glue. As support, the canvas was divided into 10×10 cm^2 dimensions. Several pigments (cinnabar, lead white, malachite, azurite, lapis lazuli, yellow ochre and Naples yellow) and binders (egg yolk, linseed oil and animal glue) were chosen for the testing. Investigations are still ongoing, and only the tests with egg yolk as a binder and cinnabar and lead white as pigments are reported here. The egg yolk was selected as a binder for its wide use in easel paintings and its fast drying compared to linseed oil. Regarding the pigments, cinnabar was selected for its known sensitivity to light, while as one of the most used white pigments, lead white [$2PbCO_3Pb(OH)_2$)] has intrinsic –OH groups in the molecule, which may increase its sensitivity to the Er:YAG laser irradiation. Therefore, considering the different interactions and possible sensitivity of pigments under laser irradiation, cinnabar and lead white were applied on the canvas with egg yolk as a binding medium to simulate the tempera painting. Table 1 illustrates the mass of materials used in mock-up preparation. All pigments were purchased from Zecchi (Florence, Italy), while eggs were freshly bought in a local supermarket in Florence. In order to prepare the painting mock-ups, egg yolk was carefully extracted from the whole egg and mixed with different pigments, and then the mixture was brushed onto canvas evenly. Typically, the mass of pigment and egg yolk was almost the same (Table 1). During the mixing, some deionized water was added to decrease the viscosity for easy application. The painted samples prepared were placed in lab conditions for 1 month to achieve complete evaporation of the water (drying). After the complete drying of the pictorial layer, solutions of one natural and one synthetic varnish, i.e., 20% mastic (tears of Chios) in turpentine oil and 44% P B67 ($\geq$98%) in acetone ($\geq$99.5%), were brushed several times until no more absorption was observed (saturation). However, due to the low concentration of mastic, the varnish was applied again on the samples 2 months later to achieve a varnish thickness of about 10–15 μm. After 6 months in the lab conditions, the samples varnished with mastic were placed in an oven at 60 °C for 10 days to accelerate the hardening/drying process of the varnished layer. Finally, the surfaces were gently touched by fingers to confirm the complete drying. The choice of these two varnishes is motivated by their extensive use since ancient times (mastic) or by their use in modern restoration interventions.

Table 1. Mass of pigments, egg yolk and water used in mock-up preparation.

Pigment	Mass of Pigment (g)	Mass of Egg Yolk (g)	Mass of Water (g)
Cinnabar	3.529	3.528	0.700
Lead white	3.589	3.960	0.800

2.2. Hyperspectral Sensor

The hyperspectral sensor employed in this study was an Analytical Spectral Devices FieldSpec FR Pro 3 (Analytical Spectral Devices, Inc., Boulder, CO, USA), a portable high-resolution spectroradiometer. It is designed to acquire Visible, Near Infrared (VNIR: 350–1000 nm) and Short-Wave Infrared (SWIR: 1000–2500 nm) punctual reflectance spectra in 0.2 s per spectrum. The sampling interval is 1.4 nm and 2 nm for the VNIR and SWIR spectral regions, respectively. The VNIR spectrometer has a standard spectral resolution (full-width at half maximum of a single emission line) of 3 nm at around 700 nm, while in the SWIR region, the spectral resolution is 10–12 nm from 1400 to 2100 nm. Spectra were firstly collected and then processed using ASD's RS3 and ViewSpec Pro 6.0 software. The instrument is equipped with a contact reflectance probe (fixed geometry of illumination and shot with a spot analysis of about 1.5 cm^2) that provides an internal light source, so spectra can be acquired without involving ambient solar illumination as in the applications of remote sensing.

The characterization of pure varnishes by the hyperspectral sensor was carried out firstly on Teflon support, on which the previously prepared Mastic and P B67 solutions were deposed to form thin films (thickness about 50–70 μm). The drying procedure of the varnish films was the same as that used for the painting mock-ups. The spectral collection was performed after the drying of varnish films. Then, the characterization of painting mock-ups without varnishes was conducted in order to serve as references when evaluating the efficacy of laser cleaning. Afterward, painting mock-ups with pictorial layers and varnishes were analyzed by the hyperspectral sensor before and after laser cleaning.

All reflectance spectra were collected with the Contact Probe by direct contact with the targeted surfaces. In order to eliminate the interferences of the ground layer and the canvas support, the background spectra were also collected on unpainted surfaces. The spectra reported were the average spectrum of 10 acquisitions performed on the same analyte. The spectra were then processed as raw by ViewSpec Pro 6.0 software.

2.3. Varnish Removal by an Er:YAG Laser

The 2.94 μm Er:YAG laser ("Light Brush" made by DEKA, El.En. Group, Calenzano, Italy), which emits 250 μs duration "macropulses" consisting of 1–2 μs micro-pulses, was used for varnish removal tests. The radiation goes directly into a 1 m long, 1 mm bore hollow internal mirrored glass guide with a pen-like tip for aiming at specific target sites. The control of the energy of each pulse to the millijoule level allows the determination of ablation thresholds for each material. For the instrument, the highest macropulse frequency is 15 Hz, and the highest energy is 100 mJ.

The laser cleaning on varnished mock-ups was operated at a relatively constant distance of 5 mm between the surface and the laser. The laser beam was delivered through the internal mirrored glass guide, and a 3–5 mm diameter of spot size was produced. Cleaning was performed without using its original flexible protecting window because the laser beam almost stopped in this case. As shown in Figure 1, each sample of 10 × 10 cm^2 dimensions was further divided into nine quadrants for evaluating different cleaning conditions. The cleaning environment was basically "dry" (without wetting agent) and "wet" (with deionized water or 2-propanol (>99.5%, Sigma-Aldrich Inc., Darmstadt, Germany) as wetting agent), and the laser irradiation conditions applied were: (1) 50 mJ, 10 Hz, 0.5 W; (2) 100 mJ, 7 Hz, 0.7 W; (3) 100 mJ, 10 Hz, 1 W for energy, frequency and power, respectively. In order to prevent the pollution of the laser beam from the cleaning substance, a square glass coverslip (1.8 × 1.8 cm^2) was placed on the cleaning area during operation. In order to clean an area of this size, 5 scans of laser ablation were taken within 150 s. During and after the cleaning, a portable hyperspectral sensor was exploited to evaluate the efficacy.

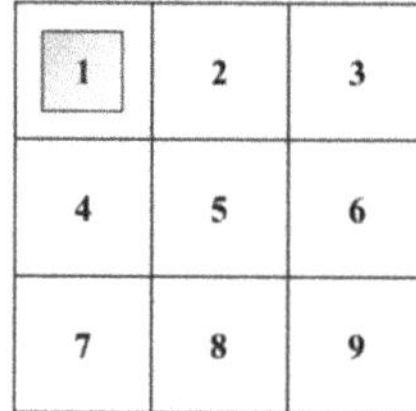

Section 1: Dry; 50 mJ, 10 Hz, 0.5 W
Section 2: Dry; 100 mJ, 10 Hz, 1 W
Section 3: Wet with H$_2$O; 100 mJ, 7 Hz, 0.7W
Section 4: Wet with H$_2$O; 50 mJ, 10 Hz, 0.5W
Section 5: Wet with H$_2$O; 100 mJ, 10 Hz, 1 W
Section 6: Wet with 2-propanol; 100 mJ, 7 Hz, 0.7W
Section 7: Wet with 2-propanol; 50 mJ, 10 Hz, 0.5W
Section 8: Wet with 2-propanol; 100 mJ, 10 Hz, 1W
Section 9: Reference

Figure 1. The different cleaning conditions operated on painting mock-ups.

2.4. Optical Microscopy

A Leica MZ 125 Stereomicroscope(Leica Microsystems, Wetzlar, Germany) equipped with a digital camera (Canon 500D, Canon Inc., Tokyo, Japan) capturing system was used to document the surface morphology of painting mock-ups before and after laser cleaning.

In order to find the same area before and after laser ablation on the surfaces of mock-ups, a paper mask with horizontal and vertical axes was used for precise locating. The position $(0, 0)$ and $(-5\ mm, -5\ mm)$ in the coordinate system for the left corner of the mock-up was selected for observation. The morphological characterization was performed under $10\times$, $32\times$ and $80\times$ of magnification.

3. Results and Discussion

3.1. Assessment of the Cleaning of Mastic

3.1.1. Cinnabar

Cinnabar (HgS) is considered an extremely light-sensitive pigment and one of the most sensitive pigments to laser irradiation [34]. Although the ablation action of the pulsed Er:YAG laser is based on the excitation of the –OH and –NH bonds due to the resonance between the radiation at 2.94 µm and the vibration stretching modes [21,35], the interaction between the laser radiation and cinnabar cannot be excluded. The results show that once a lower level of energy (50 mJ, 10 Hz) was applied in dry conditions, the superficial varnish layer was cleaned by laser ablation with no damage to the cinnabar. With the increase in power to 1 W (100 mJ, 10 Hz), the varnish layer was almost removed, but some damage to pictorial layers occurred. As shown in Figure 2b, the cinnabar turned to black spots (marked by white cycles) on which varnish was removed. The appearance of black spots instead of a uniform darkening of the irradiated area is due to the non-uniformity of the varnish layer: thicker the varnish film, less the damaging effect. Without applying wetting agents, a superficial thermal effect was induced by laser irradiation.

As the preferential absorption of the laser radiation by the –OH groups of water may prevent the thermal transformation of the pigments [35], with the presence of water, damages to pigment were not observed under the same working conditions (power 1 W, 0.7 W, 0.5 W). However, since mastic is hydrophobic, water cannot wet the varnish layer uniformly. Therefore, the cleaned surface was not uniform (Figure 2d,d′), and damages were concentrated on the areas where water was not present (marked by black circles).

It is evident that the varnish removal efficacy dramatically increased when using 2-propanol as the wetting agent (Figure 2f,f′). The surfaces (Sections 6–8) were well wetted, and the cleaning was more effective and evenly (Figure 2f). However, it is worth noticing that, even with a lower frequency (7 Hz), higher energy (100 mJ) can induce a little discoloration of cinnabar (0.7 W). The microscopic observation of Section 6 (0.7 W) is shown in Figure 2f,f′. It is clear that mastic was over-cleaned. The cleaned painting layer showed a "whitened" appearance together with some black spots.

3.1.2. Lead White

Lead white $(2PbCO_3Pb(OH)_2)$ is one of the most important white pigments since ancient times, thanks to their hiding power. It is well known, the presence of –OH groups in their crystal structure may increase its sensitivity to the Er:YAG laser irradiation. However, the results of cleaning in dry conditions showed that lead white was not damaged by laser ablation, even when the highest cleaning power (1 W, by 100 mJ and 10 Hz) was used. This is mainly due to the partial removal of the varnish, which can be deduced from the small concavities created on the surface after 150 s of pulse duration (Figure 3b).

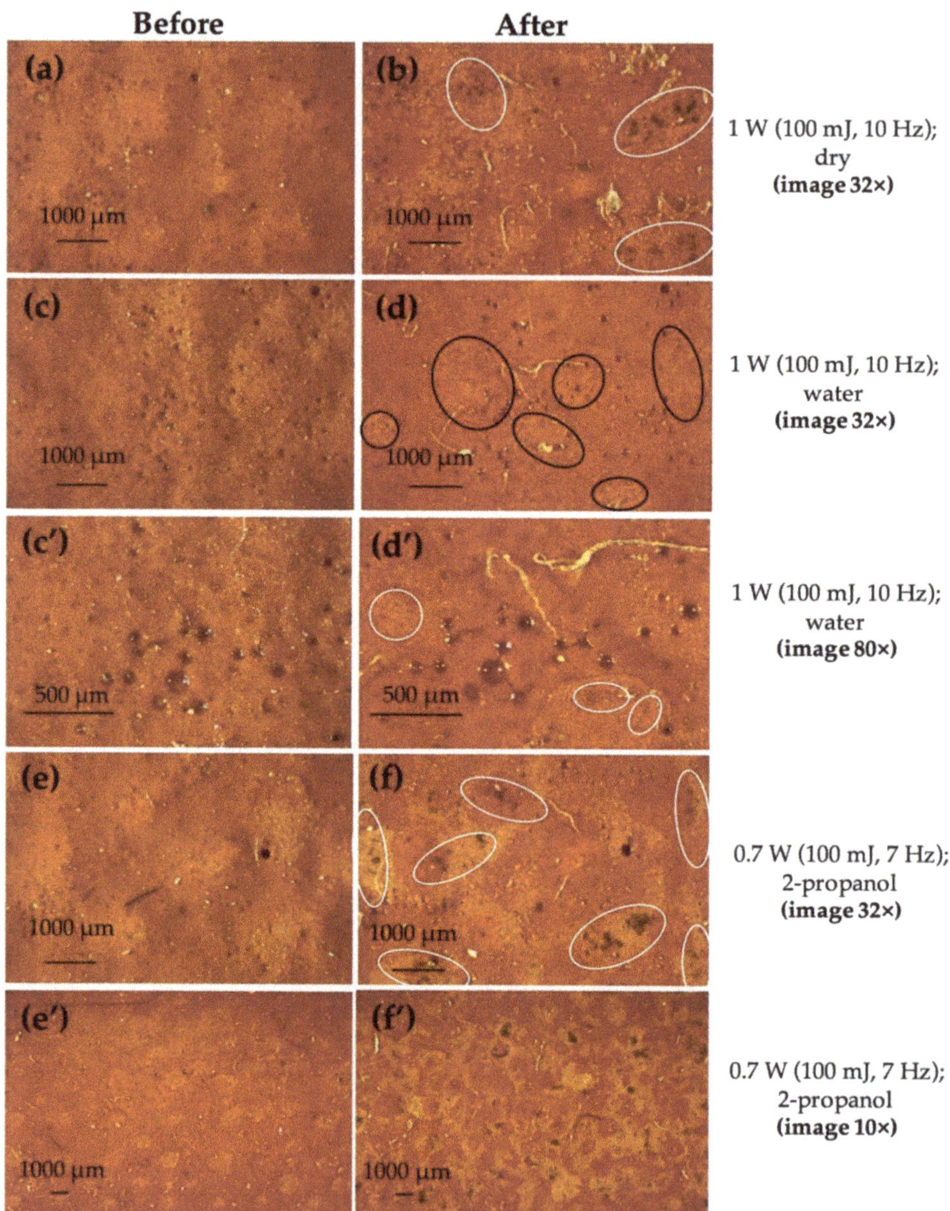

Figure 2. The microscopic images (10×, 32× and 80×) of mastic removal by Er:YAG laser under varied conditions and parameters, conducted on cinnabar tempera mock-ups. (**a,b**) is image obtained before and after cleaning in dry condition, while (**c,d**) was obtained before and after cleaning in wet (H_2O) condition. (**c′,d′**) (80×) was zoomed images of (**c,d**) at the same location, and (**e′,f′**) was taken at the same location as (**e,f**) but with 10× of magnification.

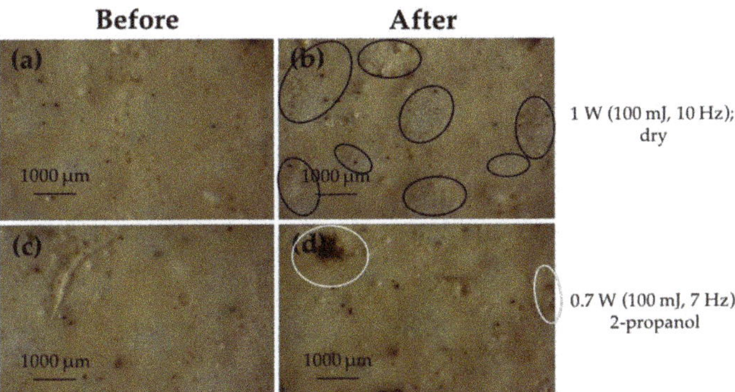

Figure 3. The microscopic images (32×) of mastic removal by Er:YAG laser under varied conditions and parameters, conducted on lead white tempera mock-ups. (**a**,**b**) was obtained before and after cleaning in dry condition with 1 W as laser power, and (**c**,**d**) is image before and after cleaning in wet (2-propanol) condition with 0.7 W as laser power.

By applying solvents as wetting agents, it was suggested that they could work as barriers to laser irradiation during cleaning [36]. This may be true when water was used as the wetting agent, as in the case of mastic varnish on the cinnabar layer. On the contrary, the surfaces (in Sections 3–5) were not cleaned at all with no varnish removal; even the energy applied was at the highest level (Figure 4). Unlike water, 2-propanol facilitated the removal of mastic already at low power (0.5 W, 50 mJ and 10 Hz), but higher power (in Sections 6 and 8) resulted in permanent discoloration—blacken of lead white (Figure 3d). Where the pigment turned black, the varnish layer was probably very thin; therefore, it was totally removed after a few seconds of these irradiation conditions, with a consequent lack of lead white protection.

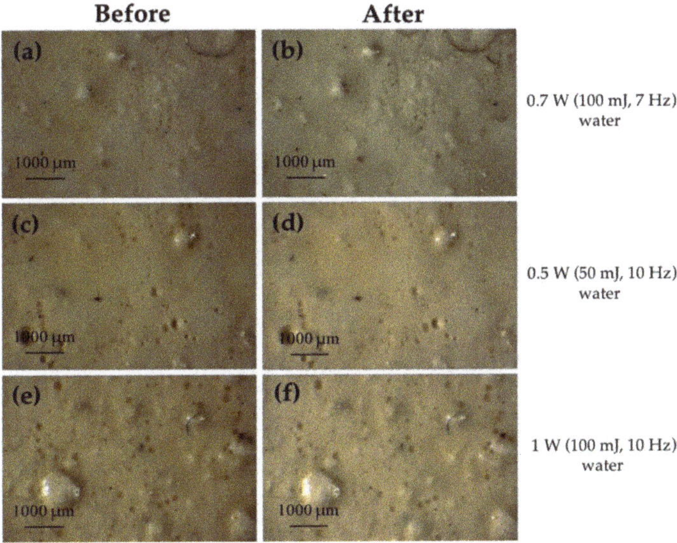

Figure 4. The microscopic images (32×) of mastic removal by Er:YAG laser under varied conditions and water as wetting agent, conducted on lead white tempera mock-ups. (**a**,**c**,**e**) was taken before cleaning, while (**b**,**d**,**f**) was obtained after cleaning in wet (H_2O) conditions with varied laser power (0.7 W, 0.5 W and 1 W respectively).

3.2. Assessment of the Cleaning of P B67

3.2.1. Cinnabar

In the case that P B67 was the varnish layer, the lowest power (0.5 W, 50 mJ, 10 Hz) did not cause any damage to the cinnabar, and no varnish was removed. By increasing the power to 1 W (100 mJ, 10 Hz), discoloration of the pigment appeared (Figure 5b), whilst the layer of P B67 exfoliated from the painting surfaces but not detached. Under this cleaning condition, the blackening of cinnabar was found, and this is mainly due to the polymorphic transformation from red hexagonal cinnabar (α-HgS) to black cubic metacinnabar (α'-HgS) [34].

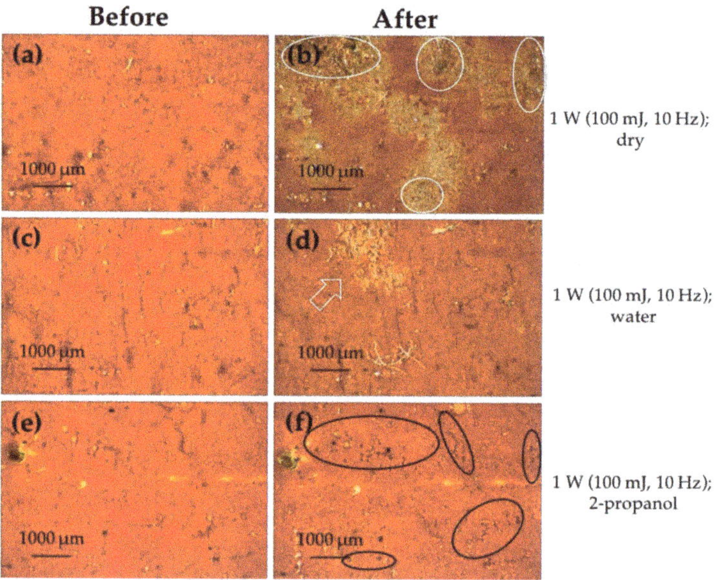

Figure 5. The microscopic images (32×) of P B67 removal by Er:YAG laser under varied conditions and parameters, conducted on cinnabar tempera mock-ups. (**a,c,e**) was obtained before cleaning, while (**b,d,f**) was obtained after cleaning in dry and wet conditions with 1 W as laser power.

When water or 2-propanol were used as a wetting agent, the temperature increase in the irradiated surface was reduced due to the presence of –OH groups in the solvents, which provide a strong absorption at 2.94 µm, ensuring the retention of much of the heat produced [22,35]. However, different from mastic, only 2-propanol provided an effective cleaning. No matter the power used, there was no more damage to the pictorial layer caused by laser irradiation. The difference between these two wet cleaning conditions was that water appeared to have no interaction with the varnish resulting in a better protecting effect than 2-propanol, and consequently, less varnish was removed. With the presence of water on the surface, even though the highest laser power (1 W, 100 mJ, 10 Hz) was adopted, P B 67 was not removed at all (no products on the covering glass), as shown in Figure 5d. Since the surface of P B67 was hydrophobic, water could not wet the entire surface evenly. Only area covered by water was well protected, while other areas were as cleaning in dry conditions, which became a layer with micro-bubbles (marked by a white arrow in Figure 5d). Similar morphological changes were also found by Chillè et al. during the cleaning of aged dammar films by using the Er:YAG laser, as they described "By increasing the fluence, the possibility of identifying differences in appearance of the spots was reduced, due to partial melting of the varnish surface and to the increase in micro-pits and bubbles" [27].

While using 2-propanol, owing to its low surface tension, 2-propanol distributed well on P B67 and led to the evenly cleaning. After laser ablation, the varnish surface became sticky, which was easy to clean with traditional cleaning methods (e.g., cotton swap) afterward. This is probably due to the fact that substances with smaller molecular weights (e.g., chain scission) were produced. With the increase in laser power, the cleaning efficacy also improved. In Figure 5f, it is evident that P B67 was removed under this laser condition without damage to the pigment.

3.2.2. Lead White

With lead white and egg yolk as the pictorial layer, P B67 varnish was cleaned using the same methods. As in the cases of mastic samples, there was no discoloration of lead white under laser ablation in dry condition, which was mainly due to the permanence of the varnish more than the stability of lead white to Er:YAG laser radiation. Figure 6a,b shows the microscopic images (32×) of pigment before and after laser cleaning. Though the highest laser power was applied (1 W), almost no varnish was removed from the surface.

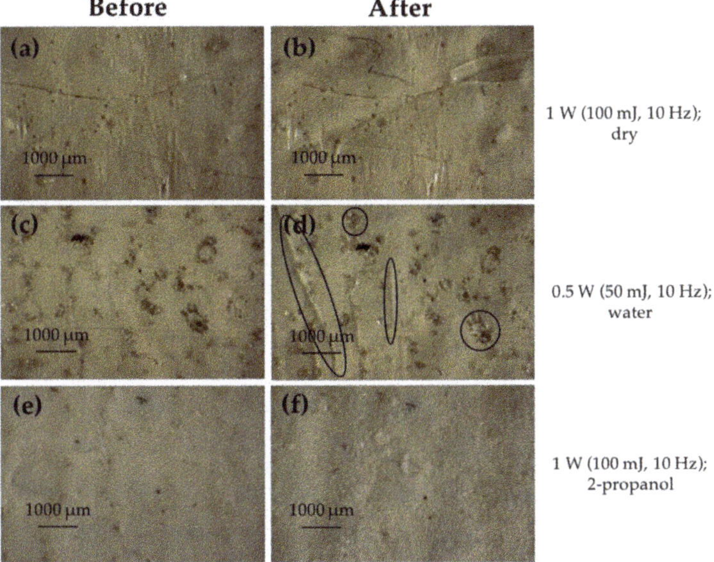

Figure 6. The microscopic images (32×) of P B67 removal by Er:YAG laser under varied conditions and parameters, conducted on lead white tempera mock-ups. (**a,c,e**) was taken before cleaning, while (**b,d,f**) was obtained after cleaning in dry and wet conditions with different laser power (1 W and 0.5 W).

When water was deposited on varnish before laser ablation, P B67 was better cleaned even with the lowest laser power. As marked by the circles in Figure 6d, line-shaped concaves formed after cleaning, indicating the partial removal of superficial varnish. The different results obtained in this case, compared with the ablation of P B67 on cinnabar, may be due to a poor homogeneity of the varnish layer, which favors a better interaction (absorption) of water with the surface. However, in this situation, water not only protected the pictorial layer but also contributed to varnish cleaning. When 2-propanol was used as a wetting agent, better cleaning efficacy was achieved. P B67 layer was cleaned better without damaging the lower layer. From Figure 6f, it was clear that P B67 became very rough with a flake-like texture after laser cleaning.

3.3. In Situ Monitoring of Cleaning by a Portable Hyperspectral Sensor

By aiming to evaluate the cleaning effectiveness of laser ablation, the reflectance spectra in full range (350–2500 nm) of the painted surface, the surface of the varnished painting and the varnished surface after laser cleaning were obtained by the rapid and in situ technique, i.e., a portable hyperspectral sensor (ASD Fieldspec FR Pro 3). The reflectance spectra collected in Section 8 of cinnabar painting mock-ups are shown in Figure 7a, and the original spectrum of pure mastic was also shown as the reference. All spectra contain cinnabar presented very typical sigmoid shape, with an inflection point characterized by a maximum peak in the 1st derivative reflectance spectrum at the wavelengths between 580 and 610 nm. Although the results of optical microscopy manifested that the cinnabar was discolored (as in Section 6, Figure 2f), the characteristic spectral features of cinnabar did not change before and after laser cleaning. In order to definitively demonstrate that the blackening of cinnabar was not induced by a chemical reaction, other techniques should be employed to detect possible new compounds formed. However, the result here found suggests a polymorphic transformation of cinnabar, which is in agreement with other research [34].

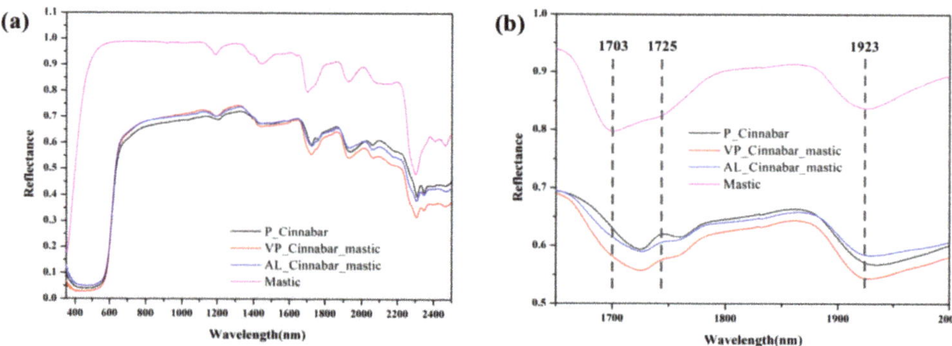

Figure 7. The full range reflectance spectrum (**a**) and the selected region reflectance spectrum (**b**) of pure mastic, cinnabar tempera mock-up (P), cinnabar tempera varnished with mastic (VP) and mock-up after laser cleaning (AL). All spectra reported were collected in Section 8 of painting mock-ups (Figure 1).

The most evident variances in the spectra collected from Section 8 varnished by mastic before and after laser cleaning (100 mJ, 10 Hz, 2-propanol as wetting agent) are plotted in Figure 7b (spectral range 1650–2000 nm). It is clear that the absorption peak of the first overtone of C–H stretching of mastic, which locates at 1703 and 1725 nm, became broader after laser cleaning [37]. In the meanwhile, the broad peak at 1923 nm attributed to the combination band of asymmetric stretching and bending of O–H bond of mastic was even broader and also shifted to a longer wavelength [37,38]. Similar modifications of spectral features were also found in the spectra of mock-ups prepared with lead white and mastic varnish (Figure 8). Regardless of the pigment involved, the reflectance spectra after laser cleaning share the same modifications, i.e., decreasing of the intensity and shifting of the characteristic absorption peaks of mastic. All the above changes make the spectra after cleaning resemble the spectrum of the painting surface without varnish.

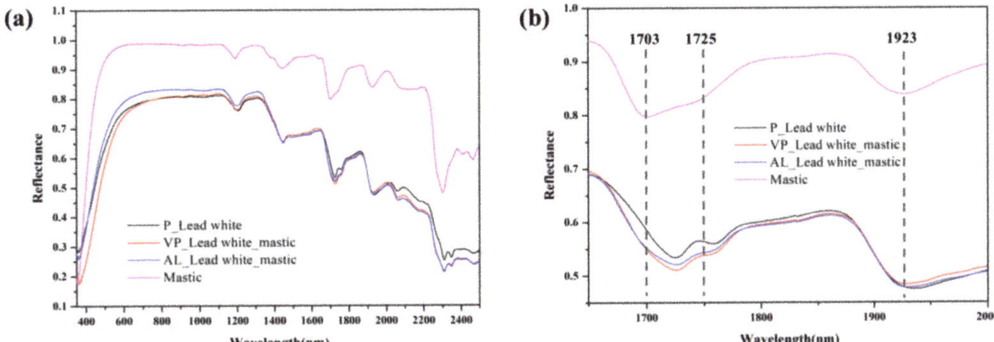

Figure 8. The full range reflectance spectrum (**a**) and the selected region reflectance spectrum (**b**) of pure mastic, lead white tempera mock-up (P), lead white tempera varnished with mastic (VP) and mock-up after laser cleaning (AL). All spectra reported were collected in Section 8 of painting mock-ups (Figure 1).

In the case of cinnabar tempera painting mock-ups varnished by P B67, we also selected Section 8 to study its reflectance spectra to monitor the efficacy of laser cleaning. In Figure 9a, it is evident that almost all the characteristic absorption peaks of P B67 can be found in the spectrum before and after cleaning. Although the cleaning was not completed, some typical features were found to be strongly weakened due to the partial removal of P B67. The absorption peak ascribed to the first overtone of C–H stretching mode is observed at around 1703, 1730 and 1762 nm for P B67, and they became less evident after cleaning. Besides, the peak intensity of 2275 nm 2308 nm corresponding to the combination band of the first overtone of C–H stretching and bending also reduced [38]. Likewise, very similar behaviors were also found in lead white tempera varnished with P B67 (Figure 10). It is noticed, under the same cleaning condition, the reduction in the intensity of the first overtone of CH_2 stretching at around 1703 nm seemed less obvious in the spectrum of P B67 compared to mastic. This can be explained by the fact the intensity of the diffuse reflectance spectrum is closely related to the roughness of the detected surface, as the surface became very rough after cleaning.

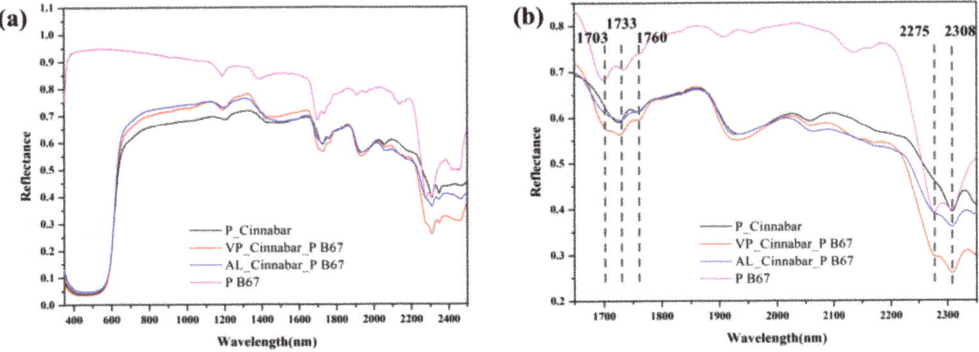

Figure 9. The full range reflectance spectrum (**a**) and the selected region reflectance spectrum (**b**) of pure P B67, cinnabar tempera mock-up (P), cinnabar tempera varnished with P B67 (VP) and mock-up after laser cleaning (AL). All spectra reported were collected in Section 8 of painting mock-ups (Figure 1).

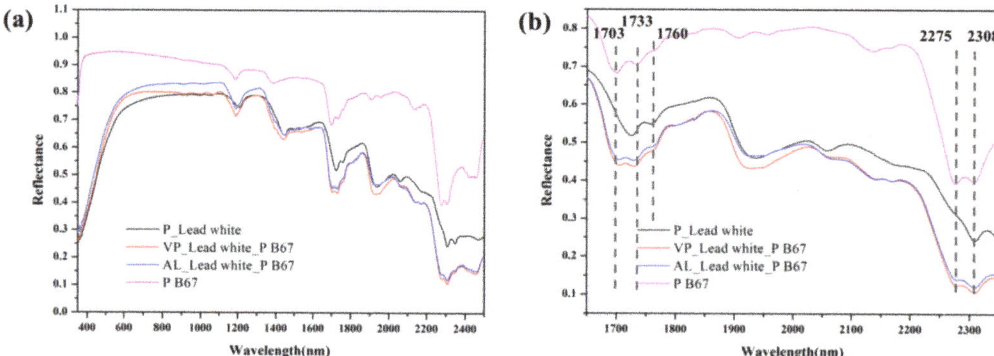

Figure 10. The full range reflectance spectrum (**a**) and the selected region reflectance spectrum (**b**) of pure P B67, lead white tempera mock-up (P), lead white tempera varnished with P B67 (VP) and mock-up after laser cleaning (AL). All spectra reported were collected in Section 8 of painting mock-ups (Figure 1).

3.4. Best Condition for Laser Sensitive Pigments

After the comparative study on cleaning results under different conditions, safe parameters should be determined for the removal of mastic and P B67 from cinnabar and lead white tempera painting through the Er:YAG laser. Based on the microscopic analysis conducted on four different combinations under various cleaning conditions (from Figures 11–14), it can be seen that cleaning efficacy increased as the power increased, while the cleaning condition (dry or wet) also strongly affected the final results depending on the type of varnish. Dry and wet cleaning conditions are always discussed by researchers when the Er:YAG laser is adopted for cleaning since wetting agents are able to decrease the temperature in the bulk and obtain a gentler interaction with varnish surface [24,25].

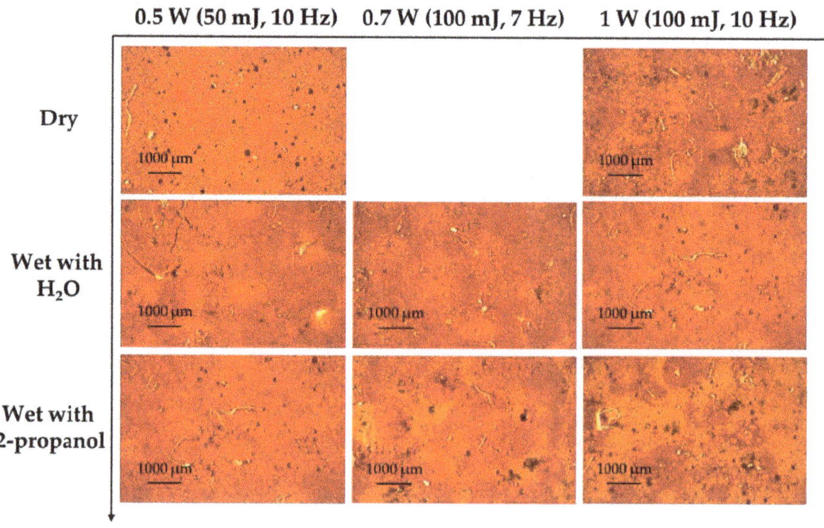

Figure 11. The microscopic images (32×) of the eight sections of mastic varnished cinnabar tempera mock-ups after different laser cleaning conditions and parameters.

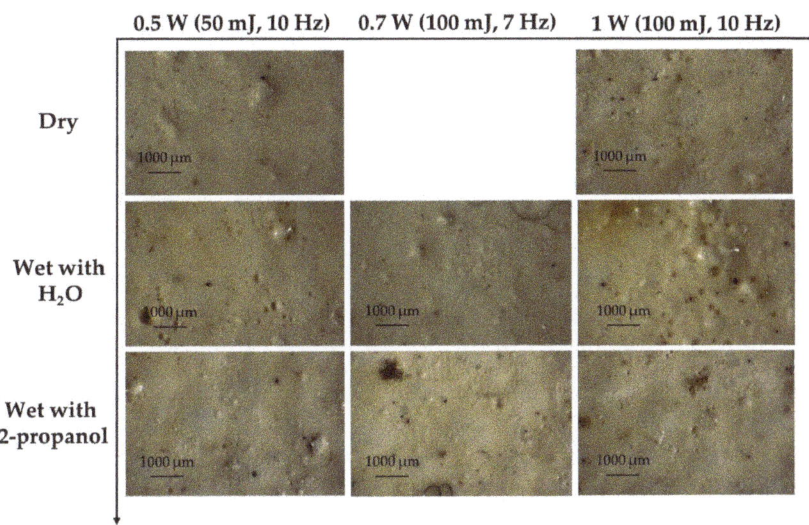

Figure 12. The microscopic images (32×) of the eight sections of mastic varnished lead white tempera mock-ups after different laser cleaning conditions and parameters.

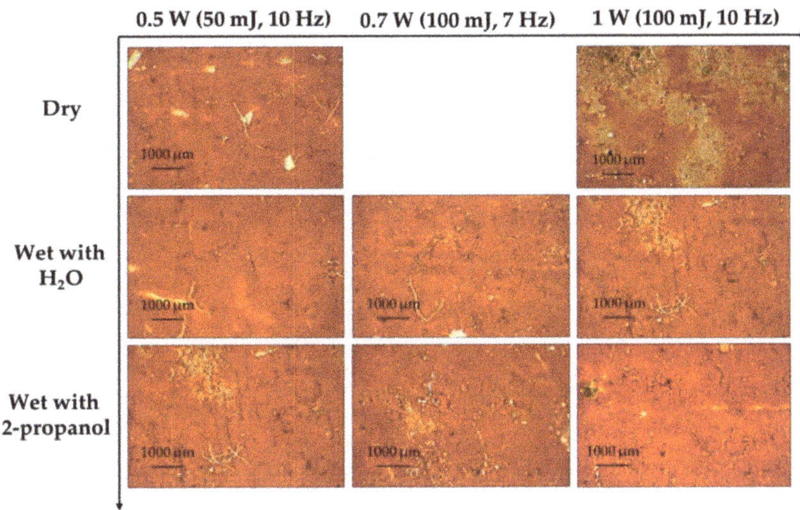

Figure 13. The microscopic images (32×) of the eight sections of P B67 varnished cinnabar tempera mock-ups after different laser cleaning conditions and parameters.

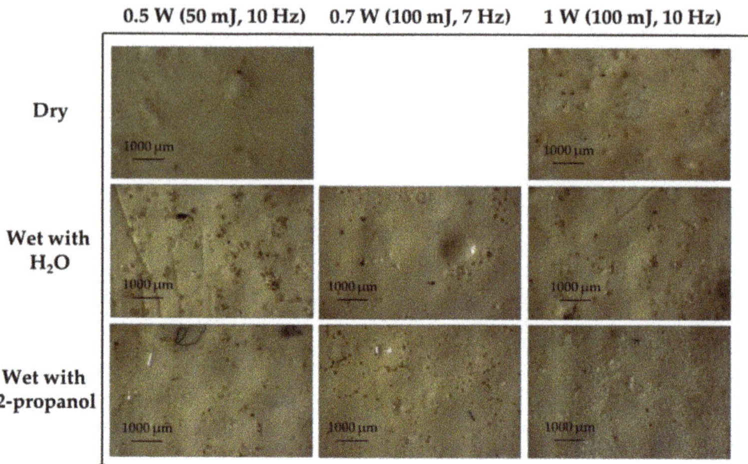

Figure 14. The microscopic images (32×) of the eight sections of P B67 varnished lead white tempera mock-ups after different laser cleaning conditions and parameters.

In the dry condition, cleaning without any wetting agents helped us determine the safety threshold for laser ablation (max power 0.5 W, by 50 mJ and 10 Hz) on light-sensitive pigments. In the wet condition, two wetting agents showed distinct influences on the cleaning efficacy. As for water, the water could not wet either the surface of mastic or P B67 evenly due to its relatively high surface tension. During laser ablation, water absorbed the major energy of laser irradiation and acted as a protective layer to the pigments in these four cleaning cases. Consequently, cinnabar and lead white were well protected, although the cleaning was less effective compared with the results obtained with the same operative parameters in dry conditions. This cleaning effect was more obvious for the more sensitive pigment—cinnabar. In Figures 11 and 13, it was clear no matter for cleaning mastic or P B67, when water was present, discoloration of cinnabar did not occur even with the highest power.

However, when 2-propanol was applied as a wetting agent, the situation became more complicated. Thanks to its proper physical–chemical properties, it is well distributed over the entire cleaning surfaces in all cases, facilitating the evenly cleaning by laser ablation. The cleaning efficacy was improved, but damages to the pictorial layer were also observed, i.e., blackening of lead white during cleaning of mastic. These results indicate the discoloration of pigments was not only related to the chemical nature of pigments but also affected by the type of varnish presented.

In the case of mastic removal, both cinnabar and lead white were damaged when high power (0.7 W and 1 W) was applied (Figures 11 and 12). Comparing the result of these 8 sections cleaned in different conditions and parameters, the worst cleaning (least varnish removal) were in Section 6 and Section 8, which implied that the two parameters (0.7 W and 1 W) with the presence of 2-propanol was not suitable for cleaning mastic varnish on cinnabar or lead white pigments. When the lowest power (0.5 W) was used, a limited amount of mastic was removed while no damage was induced to the pictorial layer. In this case, repeated cleaning can give us satisfactory results.

In the case of P B67 removal, the best results were achieved with the pre-treatment of 2-proponal before laser irradiation. For both cinnabar and lead white painting mock-ups, in wet condition, the cleaning efficacy of P B67 increased as the pigments were well protected. After applying the laser power of 0.5 W, the varnish layer became sticky and it was easy to remove without damaging the painting layer, as shown in Figures 13 and 14. Once the power of laser irradiation increased, the surface became stickier, and it was very easy to clean with the traditional method. These results implied, the thin layer of 2-propanol acted as a barrier to absorb the laser energy and limited the penetration depth of irradiation superficially simultaneously. The best cleaning effect was obtained by applying 1 W of power; p B67 was evenly and thoroughly removed on the lead white layer (Figure 14). However, possible damages may be created if a higher power was used for repeated cleaning.

4. Conclusions

In this research, the efficacy and side effects of varnish removal by a free-running Er:YAG laser at 2.94 μm were systematically investigated. By focusing on the validation and evaluation of the applicability of the Er:YAG laser for laser-sensitive pigments, cinnabar and lead white tempera mock-ups varnished with mastic and P B67 were cleaned under varied conditions and parameters. In addition to the traditional optical microscopic analysis, a novel rapid, in situ and non-invasive hyperspectral sensor was exploited to monitor the progress of varnish removal.

In general, under the same laser cleaning conditions, cinnabar was more easily damaged compared with lead white, although lead white has intrinsic O–H groups in its structure. The safety threshold of ablation energy for laser-sensitive pigments was determined as 0.5 W of power (100 mJ and 5 Hz) by tests conducted in dry conditions. Once wetting agents were introduced, they had different influences on the cleaning efficacy. Water, regardless of the pigment and varnish used, demonstrated a protective effect for the pictorial layer during cleaning, but it decreased the removal of varnish. In contrast, 2-propanol improved the cleaning efficacy due to its chemical–physical properties (e.g., lower surface tension and better wettability than water). However, 2-propanol induced the discoloration of both cinnabar and lead white in the case of mastic removal when high ablation power was applied ($\geq$0.7 W). Moreover, by providing the full range reflectance spectra, the portable hyperspectral sensor illustrated good potential in evaluating varnish cleaning as an in situ tool. The progress of varnish removal can be scientifically and rapidly controlled during laser cleaning by simply studying the spectral shape and relative peak intensity change in the spectra of varnishes and pigments.

In conclusion, with 2-propanol as a wetting agent, satisfactory cleaning can be achieved without damaging laser-sensitive pigments by adjusting the working conditions of the Er:YAG laser to 50 mJ of energy and 10 Hz of frequency (0.5 W of power).

Author Contributions: Conceptualization, M.C.; methodology, C.W. and M.C.; investigation, C.W.; writing—original draft preparation, C.W. and Y.C.; writing—review and editing, C.W., Y.C., F.T. and M.C.; supervision, M.C.; funding acquisition, C.W. and M.C. All authors have read and agreed to the published version of the manuscript.

Funding: This research was funded by the National Key Research and Development Project (Grant No. 2020YFC1521904), the Social Science Foundation of Shaanxi Province (Grant NO. 2021G011), the Scientific Research Program of Shaanxi Provincial Education Department (Grant NO. 21JK0947) and the start-up funding from Northwest University, China.

Institutional Review Board Statement: Not applicable.

Informed Consent Statement: Not applicable.

Data Availability Statement: Not applicable.

Acknowledgments: The authors thank El. En. Group for supplying the Er:YAG laser source. Sandro Moretti and Teresa Salvatici from Department of Earth Sciences-University of Florence are also acknowledged for their help with the application of the hyperspectral sensor.

Conflicts of Interest: The authors declare no conflict of interest.

References

1. De la Rie, E.R. The inpower of varnish on the appearance of paintings. *Stud. Conserv.* **1987**, *32*, 1–13.
2. Reifsnyder, J.M. A note on a traditional technique of varnish application for paintings on panel. *Stud. Conserv.* **1996**, *41*, 120–122.

3. Phenix, A.; Sutherland, K. The cleaning of paintings: Effects of organic solvents on oil paint films. *Stud. Conserv.* **2001**, *46*, 47–60. [CrossRef]

4. Baij, L.; Hermans, J.; Ormsby, B.; Noble, P.; Iedema, P.; Keune, K. A review of solvent action on oil paint. *Herit. Sci.* **2020**, *8*, 43. [CrossRef]

5. Khandekar, N. A survey of the conservation literature relating to the development of aqueous gel cleaning on painted and varnished surfaces. *Stud. Conserv.* **2000**, *45*, 10–20. [CrossRef]

6. Erhardt, D.; Tsang, J.S. The extraction components of oil paint films. *Stud. Conserv.* **1990**, *45*, 93–97. [CrossRef]

7. Carretti, E.; Dei, L.; Weiss, R.G.; Baglioni, P. A new class of gels for the conservation of painted surfaces. *J. Cult. Herit.* **2008**, *9*, 386–393. [CrossRef]

8. Mastrangelo, R.; Chelazzi, D.; Poggi, G.; Fratini, E.; Buemi, L.P.; Petruzzellis, M.L.; Baglioni, P. Twin-chain polymer hydrogels based on poly (vinyl alcohol) as new advanced tool for the cleaning of modern and contemporary art. *Proc. Natl. Acad. Sci. USA* **2020**, *117*, 7011–7020. [CrossRef] [PubMed]

9. Rui, B.; Morais, P.J.; Helena, G.; Young, C. Laser Cleaning of easel paintings: An overview. *Laser Chem.* **2006**, *2006*, 90279.

10. Ganeev, R. Laser Cleaning of Art. In *Laser—Surface Interactions*, 1st ed.; Springer: Dordrecht, The Netherlands, 2014; pp. 87–103.

11. Asmus, J.F.; Guattari, G.; Lazzarini, L.; Musumeci, G.; Wuerker, R.F. Holography in the conservation of statuary. *Stud. Conserv.* **1973**, *18*, 49–63.

12. Siano, S.; Agresti, J.; Cacciari, I.; Ciofini, D.; Mascalachi, M.; Osticioli, I.; Mencaglia, A. Laser cleaning in conservation of stone, metal, and painted artifacts: State of the art and new insights on the use of the Nd:YAG lasers. *Appl. Phys.* **2012**, *106*, 419–446. [CrossRef]

13. Siano, S.; Giamello, M.; Bartoli, L.; Mencaglia, A.; Parfenov, V.; Salimbeni, R. Laser cleaning of stone by different laser pulse duration and wavelength. *Laser Phys.* **2008**, *18*, 27–36. [CrossRef]

14. Osticioli, I.; Mascalchi, M.; Pinna, D.; Siano, S. Removal of verrucaria nigrescens from carrara marble artefacts using Nd:YAG lasers: Comparison among different pulse durations and wavelengths. *Appl. Phys. A* **2015**, *118*, 1517–1526. [CrossRef]

15. Koh, Y.S.; Sárady, I. Cleaning of corroded iron artefacts using pulsed TEA CO_2- and Nd:YAG-lasers. *J. Cult. Herit.* **2003**, *4*, 129–133. [CrossRef]

16. Sansonetti, A.; Colella, M.; Letardi, P.; Salvadori, B.; Striova, J. Laser cleaning of a nineteenth-century bronze sculpture: In situ multi-analytical evaluation. *Stud. Conserv.* **2015**, *60*, 28–33. [CrossRef]

17. Bilmes, G.M.; Vallejo, J.; Costa Vera, C.; Garcia, M.E. High efficiencies for laser cleaning of glassware irradiated from the back: Application to glassware historical objects. *Appl. Phys. A* **2018**, *124*, 1–11. [CrossRef]

18. Gaetani, C.; Santamaria, U. The laser cleaning of wall paintings. *J. Cult. Herit.* **2000**, *1*, S199–S207. [CrossRef]

19. Andreotti, A.; Colombini, M.P.; Nevin, A.; Melessanaki, K.; Pouli, P.; Fotakis, C. Multianalytical study of laser pulse duration effects in the IR laser cleaning of wall paintings from the monumental cemetery of Pisa. *Laser Chem.* **2006**, *2006*, 39046. [CrossRef]

20. De Cruz, A.; Wolbarsht, M.L.; Hauger, S.A. Laser removal of contaminants from painted surfaces. *J. Cult. Herit.* **2000**, *1*, 173–180. [CrossRef]

21. Teppo, E. Introduction: Er:YAG lasers in the conservation of artworks. *J. Inst. Conserv.* **2020**, *43*, 2–11. [CrossRef]

22. De Cruz, A.; Andreotti, A.; Ceccarini, A.; Colombini, M.P. Laser cleaning of works of art: Evaluation of the thermal stress induced by Er:YAG laser. *Appl. Phys. B* **2014**, *117*, 533–541. [CrossRef]

23. Bracco, P.; Lanterna, G.; Matteini, M.; Nakahara, K.; Colombini, M.P. Er:YAG laser: An innovative tool for controlled cleaning of old paintings: Testing and evaluation. *J. Cult. Herit.* **2003**, *4*, 202–208. [CrossRef]

24. Striova, J.; Salvadori, B.; Fontana, R.; Sansonetti, A.; Barucci, M.; Pampaloni, E.; Marconi, E.; Pezzati, L.; Colombini, M.P. Optical and spectroscopic tools for evaluating Er:YAG laser removal of shellac varnish. *Stud. Conserv.* **2015**, *60*, 91–96. [CrossRef]

25. Chillè, C.; Papadakis, V.M.; Theodorakopoulos, C. An analytical evaluation of Er:YAG laser cleaning tests on a nineteenth century varnished painting. *Microchem. J.* **2020**, *158*, 105086. [CrossRef]

26. Brunetto, A.; Bono, G.; Frezzato, F. Er:YAG laser cleaning of 'San Marziale in Gloria' by Jacopo Tintoretto in the Church of San Marziale, Venice. *J. Inst. Conserv.* **2020**, *43*, 44–58. [CrossRef]

27. Chillè, C.; Sala, F.; Wu, Q.; Theodorakopoulos, C. A study on the heat distribution and oxidative modification of aged dammar films upon Er:YAG laser irradiation. *J. Inst. Conserv.* **2020**, *43*, 59–78. [CrossRef]

28. Pereira-Pardo, L.; Melita, L.N.; Korenberg, C. Tackling conservation challenges using erbium lasers: Case studies at the British Museum. *J. Inst. Conserv.* **2020**, *43*, 25–43. [CrossRef]

29. Striova, J.; Fontana, R.; Barbetti, I.; Pezzati, L.; Fedele, A.; Riminesi, C. Multisensorial assessment of laser effects on shellac applied on wall paintings. *Sensors* **2021**, *21*, 3354. [CrossRef]

30. Camaiti, M.; Vettori, S.; Benvenuti, M.; Chiarantini, L.; Costagliola, P.; Di Benedetto, F.; Moretti, S.; Paba, F.; Pecchioni, E. Hyperspectral sensor for gypsum detection on monumental buildings. *J. Geophys. Eng.* **2011**, *8*, S126–S131. [CrossRef]

31. Wang, C.; Salvatici, T.; Camaiti, M.; Chiara, D.; Moretti, S. A new application of hyperspectral radiometry: The characterization of painted surfaces. In Proceedings of the EGU General Assembly 2016, Vienna, Austria, 23–29 April 2016.

32. Wang, C. Hyperspectral Sensor: A New Approach for Evaluating the Efficacy of Laser Cleaning in the Removal of Varnishes and Overpaintings. Master's Thesis, University of Bologna, Bologna, Italy, 2015.

33. Camaiti, M.; Benvenuti, M.; Costagliola, P.; Benedetto, F.D.; Moretti, S. Hyperspectral sensors for the characterization of cultural heritage surfaces. In *Sensing the Past*; Masini, N., Soldovieri, F., Eds.; Springer International Publishing: Gewerbestrasse, Switzerland, 2017; Volume 16, pp. 289–311.
34. Pouli, P.; Emmony, D.C.; Madden, C.E.; Sutherland, I. Studies towards a thorough understanding of the laser-induced discoloration mechanisms of medieval pigments. *J. Cult. Herit.* **2003**, *4*, 271s–275s. [CrossRef]
35. Striova, J.; Camaiti, M.; Castellucci, E.M.; Sansonetti, A. Chemical, morphological and chromatic behavior of mural paintings under er:yag laser irradiation. *Appl. Phys. A* **2011**, *104*, 649–660. [CrossRef]
36. Camaiti, M.; Matteini, M.; Sansonetti, A.; Striová, J.; Castelluci, E.; Andreotti, A.; Colombini, M.P.; De Cruze, A.; Palmer, R. The Interaction of Laser Radiation at 2.94 μm with Azurite and Malachite Pigments. Lasers in the Conservation of Artworks. In Proceedings of the International Conference Lacona VII, Madrid, Spain, 17–21 September 2008; pp. 253–258.
37. Invernizzi, C.; Rovetta, T.; Licchelli, M.; Malagodi, M. Mid and near-infrared reflection spectral database of natural organic materials in the cultural heritage field. *Int. J. Anal. Chem.* **2018**, *2018*, 7823248. [CrossRef] [PubMed]
38. Schwanninger, M.; Rodrigues, J.C.; Fackler, K. A review of band assignments in near infrared spectra of wood and wood components. *J. Near Infrared Spectrosc.* **2011**, *19*, 287–308. [CrossRef]

Article

Identification of Colourants and Varnishes in a 14th Century Decorated Wood-Carved Door of the Dionysiou Monastery in Mount Athos

Alexander Konstantas [1], Ioannis Karapanagiotis [2] and Stamatis C. Boyatzis [1,*]

[1] Department of Conservation of Antiquities and Works of Art, University of West Attica, 12243 Athens, Greece; alex.konstanta@gmail.com

[2] Department of Management and Conservation of Ecclesiastical Cultural Heritage Objects, University Ecclesiastical Academy of Thessaloniki, 54250 Thessaloniki, Greece; y.karapanagiotis@aeath.gr

* Correspondence: sboyatzis@uniwa.gr

Abstract: A decorated and carved wooden door of the late Byzantine period (14th Century), which belongs to the Dionysiou Monastery in Mount Athos, Greece, constitutes an important relic of valuable technological information due to its construction technology and history. Seventeen (17) samples detached from the door are studied using optical microscopy, scanning electron microscopy (SEM) with energy dispersive X-ray analysis (SEM-EDX), and micro-Raman and FTIR spectroscopy. The following materials are identified in the cross sections of the door samples using micro-Raman spectroscopy: orpiment, lead white, red lead, red ochre, cinnabar, carbon black, gypsum, anhydrite, and calcite, and an organic colourant of the indigoid family. SEM-EDX studies supported to the aforementioned Raman results. Interestingly, a combination of inorganic and organic colourants was detected. The main goals of this particular study were to: (a) reveal the colour palette and materials, (b) identify the type of varnish and its condition, and (c) contribute to future restoration processes and aid conservators in selecting compatible restoration materials.

Keywords: 14th century; wood-carved; microscopy; Raman; SEM-EDX; FTIR; materials; pigments

Citation: Konstantas, A.; Karapanagiotis, I.; Boyatzis, S.C. Identification of Colourants and Varnishes in a 14th Century Decorated Wood-Carved Door of the Dionysiou Monastery in Mount Athos. *Coatings* **2021**, *11*, 1087. https://doi.org/10.3390/coatings11091087

Academic Editor: Alessandro Pezzella

Received: 31 July 2021
Accepted: 27 August 2021
Published: 8 September 2021

Publisher's Note: MDPI stays neutral with regard to jurisdictional claims in published maps and institutional affiliations.

1. Introduction

A decorated wood-carved door, in functional condition, located in the Church of Dionysiou Monastery of Mount Athos, has received attention due to its historical importance. According to sources [1], the door dates back to the 14th century, thus corresponding to the first period of the history of the Dionysiou Monastery. In this period, the monastery suffered from a severe fire incident, and the decorated door is among the few surviving objects from that first period. Moreover, the door panel is covered with a metal leaf on which a coloured, carved decoration, a rare and excellent example of Byzantine wood carving, is mounted.

The origins of the decoration's colourants/pigments from this particular period have only been briefly studied on some selected icons and relics [2–5]. The goal of the present report was to document and qualitatively analyse the materials used for the decoration of this door by applying microscopy and elemental and molecular analytical techniques, and also to assess their current condition as a result of past historical events. For this reason, seventeen (17) samples (sampling spots shown in Figure 1) were sampled and investigated. Attention was focused on the pigments, the varnish coating condition, and the related stratigraphy that could reveal more possible interventions or repairing/retouching actions.

The goals of the investigation were achieved using the following methods: (i) micro-Raman and SEM-EDX were employed to identify the pigments; (ii) FTIR was used to study the organic materials of the varnish layer; (iii) and finally, optical microscopy was applied to investigate sample stratigraphies. Consequently, the final goal of the study was

61

to reveal the combinations of inorganic and organic materials, and, where possible, also their condition.

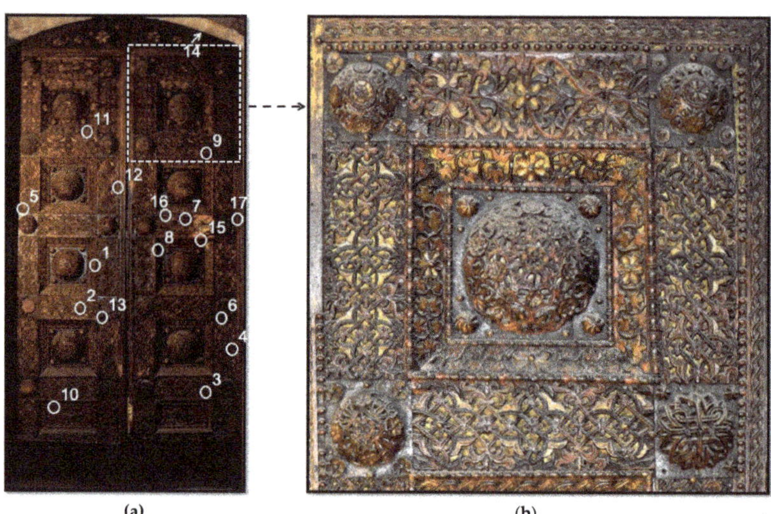

Figure 1. (**a**) Sampling spots and (**b**) detail from the decorated wood-carved door of the Church of Dionysiou Monastery in Mount Athos. Dimensions of each door leaf: 240 cm × 60 cm.

2. Materials and Methods

Aiming to investigate the apparently variable stratigraphy of the object, fifteen of the seventeen microsamples were taken from the artwork (sampling points shown in Figure 1), embedded in transparent polyester resin (Struers), and subsequently ground and polished using a Struers LaboPol-5 polishing, rotating table. The cross sections of the samples were then studied using optical microscopy, Raman and FTIR spectroscopies, and SEM-EDX scanning electron microscopy (SEM) with energy dispersive X-Ray analysis. Another two microsamples, which were scraps from the varnish layer, were extracted from the relic and subjected to FTIR analysis (samples 13 and 16).

Microsamples were studied using a Zeiss Axioskop 40 polarising light microscope equipped with a UV source, and by scanning electron microscopy (SEM; JEOL, JSM-6510, Tokyo, Japan) with energy dispersive X-ray spectroscopy (EDX; Swift-ED, Oxford Instruments, High Wycombe, United Kingdom). For the SEM and SEM-EDX studies, samples were coated with a thin layer of carbon.

The micro-Raman spectra were obtained using a Renishaw Raman device (Renishaw, London, England) equipped with a microscope and a CCD detector. As a source of excitation, a diode laser with a radiant emission wavelength of $\lambda = 785$ nm was used. The beam size of the laser was of the order of ~2 μm. The power density on the surface of the sample was ~0.5 mW/μm^2 and the spectral resolution was ~4 cm^{-1}. Alignment and calibration of the spectrometer, before and after each measurement, was achieved by using a crystalline silicon (Si) wafer.

Finally, the FTIR spectroscopy analyses were conducted so as to provide enhanced information. In particular, a Perkin-Elmer® model Spectrum 400 FT-IR spectrometer (PerkinElmer, Inc., Waltham, MA, USA) operating at the medium and near-infrared range was used. The spectra were collected with a spectral resolution of 4 cm^{-1}, in the spectral range of 4000–450 cm^{-1} and 256 scans. The analysis was performed in diffuse reflectance mode with Perkin Elmer's special 'KBr pad' design involving a standard KBr disc holding the powder sample on its surface.

3. Results

3.1. Cross Section Analysis

The investigation of the sample cross sections with optical microscopy revealed the various layers and provided the background for the subsequent spectroscopic study. In certain areas of the door, later interventions (overpaints) were revealed. The most interesting result was recorded in the samples D14 and D17 (shown in Figure 2 in visible-reflected and UV-fluorescence light, respectively), as two successive ground and painting layers (3,4 and 2,3 in Figure 2a–d, respectively) beyond the original painting layers (2 and 1 in Figure 2a–d, respectively) were observed in the microphotographs.

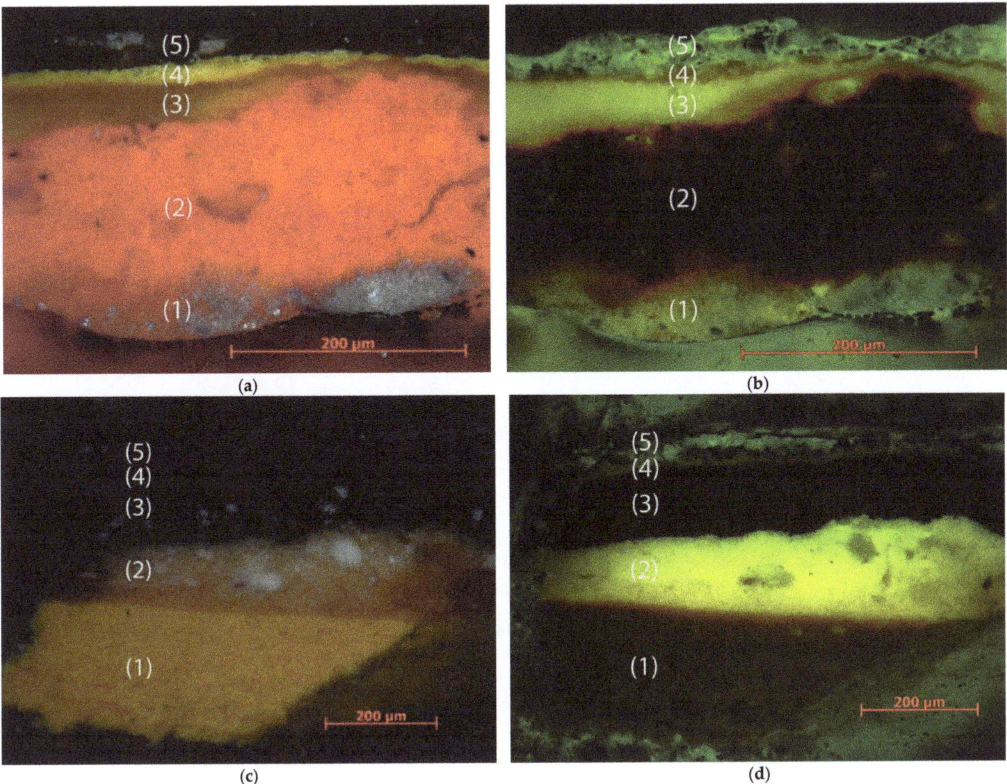

Figure 2. Sample cross sections D17 (**a**,**b**) and D14 (**c**,**d**). Cross sections in (**a**,**c**) visible-reflected light and (**b**,**d**) UV fluorescence (λexc = 365 nm) are shown. Complete stratigraphy for (**a**) and (**b**) is as follows: (1) first layer of gesso ground: gypsum; (2) first painting layer: cinnabar and grains of carbon black; (3) second layer of gesso ground: gypsum; (4) second painting layer: orpiment; (5) varnish. For (**c**) and (**d**): (1) first painting layer: orpiment; (2) layer of gesso ground: gypsum, calcite, and orpiment; (3) second painting layer: carbon black; (4) first varnish layer; (5) second varnish layer.

As the varnish layer between the overpainted layer and the original layer (layer (2) and layer (3) in Figure 2a,b and layer (1) and (2) in Figure 2c,d) is not observed, it is concluded that the overpainted area corresponds to the same period of time and the same artist, and it is probably a redaction of the artist's final work.

The results of the microscopic, Raman, and EDX analyses are summarized in Table 1, which reveals the materials used in the layers of the sample cross sections. The extensive use of lead white is shown in the data of Table 1.

Table 1. Cross sections of the samples. Pigments were identified using micro-Raman spectroscopy and SEM-EDX.

Sample/Sampling Point	Stratigraphy	Raman Results	EDX Results
D 1-Metal leaf from decoration	Back side	-	Cu, Zn
	Front side	-	Cu, Zn, Au
D 2-Yellow spot	1st painting layer	Orpiment	As, S, C, O
	1st varnish layer	-	-
	Soot layer	Carbon black	C, O
	2nd varnish layer	-	-
D 3-Red spot	Gesso ground	Gypsum	Ca, S, C, O
	1st painting layer	Lead white	Pb, O, C
	2nd painting layer	Minium and lead white	Pb, O, C
	Varnish layer	-	-
D 4-Light blue spot	1st painting layer	Indigo/woad, lead white, red ochre, calcite, and gypsum	Ca, S, Pb, C, O
	2nd painting layer	Lead white, indigo/woad, red ochre, calcite, and gypsum	Ca, S, Pb, C, O
	Varnish layer	-	-
D 5-Light blue spot	1st painting layer	Lead white	Pb, Cu, C, O
	2nd painting layer	Indigo/woad, lead white, red lead, and calcite	Pb, Cu, Ca, C, O
	3rd painting layer	Indigo/woad, lead white and calcite	
	1st varnish layer	-	-
	Soot layer	Carbon black	C, O
	2nd varnish layer	-	-
D 6-Blue spot	Gesso ground	Gypsum	Ca, S, C, O
	1st painting layer	Lead white	Pb, C, O
	2nd painting layer	Indigo/woad, gypsum, and lead white	Pb, Ca, S, C, O
	1st varnish layer	-	-
	2nd varnish layer	-	-
D 7-Green spot	1st painting layer	Lead white	Cu, Pb, C, O
	1st varnish layer	-	-
	Soot layer	Carbon black	C, O
	2nd varnish layer	-	-
D 8-Light red spot	1st painting layer	Red lead and lead white	Pb, C, O
	1st varnish layer	-	-
	Soot layer	Carbon black	C, O
	2nd varnish layer	-	-
D 9-Light blue spot	1st painting layer	Lead white	Pb, Cu, C, O
	2nd painting layer	Lead white	Pb, Cu, C, O
	1st varnish layer	-	-
	Soot layer	Carbon black	C, O
	2nd varnish layer	-	-

Table 1. *Cont.*

Sample/Sampling Point	Stratigraphy	Raman Results	EDX Results
D 10-Red spot	Gesso ground	Gypsum	Ca, S, C, O
	1st painting layer	Lead white	Pb, C, O
	2nd painting layer	Red lead	Pb, C, O
	1st varnish layer	-	-
	Soot layer	Carbon black	C, O
	2nd varnish layer	-	-
D 11-Light blue spot	1st painting layer	Red lead	Pb, C, O
	2nd painting layer	Indigo/woad and lead white	Pb, C, O
	Varnish layer	-	-
D12-Red spot	Gesso ground	Gypsum	Ca, S, C, O
	1st painting layer	Cinnabar	Hg, S, C, O
	Varnish layer	-	-
D 14-Black spot	1st painting layer	Orpiment	As, S, C, O
	Gesso ground	Gypsum, calcite, and orpiment	Ca, S, As, C, O
	2nd painting layer	Carbon black	C, O
	1st varnish layer	-	-
	Soot layer	Carbon black	C, O
	2nd varnish layer	-	-
D 15-Green spot	Gesso ground:	Gypsum and anhydrite	Ca, S, C, O
	1st painting layer	Orpiment, indigo/woad, and anhydrite	Ca, S, As, C, O
	Varnish layer	-	-
D 17-Yellow spot	1st layer of gesso ground	Gypsum	Ca, S, C, O
	1st painting layer	Cinnabar and carbon black	Hg, S, C, O
	2nd layer of gesso ground	Gypsum	Ca, S, C, O
	2nd painting layer	Orpiment	As, S, C, O
	Varnish layer	-	-

3.2. SEM-EDX Results

Figures 3 and 4 are provided as examples of the SEM-EDX results. The identification of the elements in the designated spot of the SEM image is provided. The same analysis was carried out for all samples included in the study [6–12].

The detection of Cu and Zn in sample D1 (Figure 3) is indicative of a metal alloy [6] containing the above metals, where the small amount of gold (Au) suggests that the particular metal leaf was gold-gilded (gold plated brass) [7]. The detection of As and S in sample D2 (Figure 4) shows the use of orpiment pigment by the artist [8,9]. Moreover, in samples D5, D7, and D9, the green pigment granules were investigated by SEM-EDX analysis, where copper was identified. Since luminescence prevented efficient Raman analysis in these pigments, further study is needed for more detailed identification.

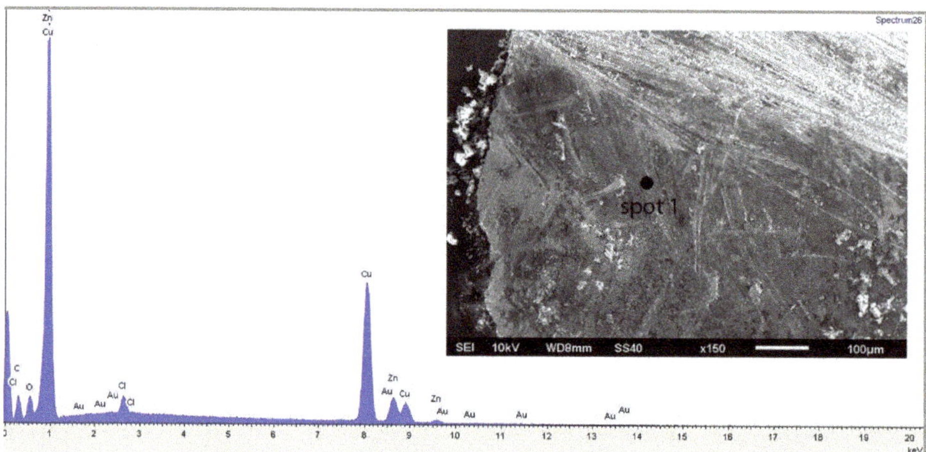

Figure 3. SEM image of the sample D1 and EDX spectrum of spot 1 in sample D1. Cu, Zn, and Au were identified.

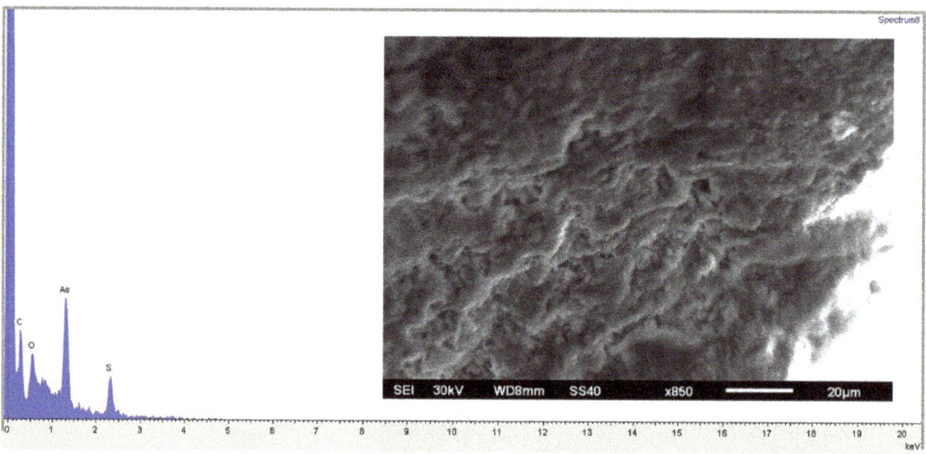

Figure 4. SEM image of the sample D2 and EDX spectrum. Elements of As and S were identified and confirm the Raman results, which revealed the presence of orpiment (As_2S_3).

3.3. Raman Results

Colourants and gesso ground materials identified in cross sections using micro-Raman spectroscopy are summarized in Table 2. Assignments of Raman peaks detected in the spectra are described next.

The Raman spectra of cinnabar displayed the characteristic strong band at ~254 cm^{-1} as well as the two weak peaks at ~284 and ~43 cm^{-1}, which were attributed to the stretching vibrations of the Hg–S bond [13–15].

In the Raman spectra of red ochre (hematite), the phonon modes of the pigment's octahedral structure at ~225, 295, 408, 498, and 611 cm^{-1} were observed, as well as the broad high-energy overtone line at ~1320 cm^{-1} [14,16].

Cross section analyses of samples D3 and D11 revealed the presence of red grains that consist of minium (Pb_3O_4), according to the characteristic predominant Raman bands at ~480 and 550 cm^{-1}, which are attributed to the stretching vibrations of the Pb–O bonds of the PbO_6 octahedra that are formed along the main symmetry axis of the structure [9]. Furthermore, three low-energy weak modes at ~232, 314 and 391 cm^{-1} were recorded in the Raman spectrum, which involve vibrations of longer Pb–O bonds interconnected by the above-mentioned chains of octahedra [17].

Table 2. Colourants identified in the investigated samples using micro-Raman spectroscopy. Characteristic peaks used for the identifications are included.

Pigment	Sample	Composition	Raman Characteristic Peaks and Relative Intensities [1] (cm^{-1})
Orpiment	D1, D14, D15, D17	As_2S_3	105 m, 136 s, 154 s, 179 m, 202 s, 293 s, 311 vs, 355 vs, 383 m
Lead white	D3, D4, D5, D6, D7, D8, D9, D10, D11	$2PbCO_3 \cdot Pb(OH)_2$	1055 vs
Minium	D3	Pb_3O_4	232 w, 314 w, 391 w, 480 s, 550 vs
Red ochre	D4	Fe_2O_3	225 vs, 295 s, 408 m, 498 w, 661 vw, 1320 m and br
Cinnabar	D12, D17	HgS	254 vs, 285 w, 343 m
Indigo/woad	D4, D5, D6, D11, D15	$C_{16}H_{10}N_2O_2$	547 w, 600 w, 942 w, 1015 w, 1147 m, 1253 s, 1313 w, 1366 vs, 1463 m, 1486 m, 1573 s, 1585 s, 1616 s, 1700 s
Carbon black	D2, D5, D7, D8, D9, D10, D14, D17	C	1318 s and br, 1584 vs and br
Gypsum	D3, D4, D6, D10, D12, D14, D15, D17	$CaSO_4 \cdot 2H_2O$	1008 vs
Anhydrite	D15	$CaSO_4$	1016 vs
Calcite	D4, D5, D14	$CaCO_3$	155 w, 280 w, 1088 vs

[1] vs: very strong, s: strong, m: medium, w: weak, vw: very weak, br: broad, sh: shoulder.

The results in Table 1 reveal the extensive use of lead white; this pigment was identified in several samples. Lead white ($2PbCO_3 \cdot Pb(OH)_2$) was identified through the characteristic peak at 1055 cm^{-1} in its Raman spectrum due to the v_1, or the symmetric stretch vibration of CO_3^{2-} [14].

Small black areas (granules) were revealed in the cross sections of several samples, which consisted of carbon black as observed in the appearance of a characteristic broad band at ~1584 cm^{-1}, assigned to a doubly degenerate deformation vibration of the aromatic ring and a second one at ~1318 cm^{-1}, and attributed to a crystalline size effect [18,19].

Apart from the inorganic pigments described above, an organic blue indigoid colourant of plant origin was also detected, which can either be correlated with indigo (*Indigofera tinctoria* L. and other species) or woad (*Isatis tinctoria* L.). Its Raman spectrum (Table 1) showed the two weak, low-energy peaks at ~547 and 600 cm^{-1} that were assigned to the bending vibrational modes involving the central [-C=C-] coordinate [20]; a weak ~942 cm^{-1} band due to the C–H out-of-plane bending motion; the peaks at ~1015 and 1147 cm^{-1}, attributed to vibrations involving the C–H in-plane bending; the 1313, 1463, 1486, 1573, and 1616 cm^{-1} spectral lines due to the C–C stretching vibrations of the benzene rings; the band at 1585 cm^{-1}, attributed to the in-phase stretching vibrations of the aromatic C=C bonds; and finally, the considerably stronger three modes at ~1253, 1366, and 1700 cm^{-1}, which were assigned to C–H/C=O bending, N–H/C–H in-plane bending, and C=O/C=C stretching vibrations, respectively [20,21].

In several samples, calcite was detected within the painting layers, possibly as a filler/dispersing medium; its Raman spectrum showed the predominant band at ~1088 cm^{-1}, which was due to the v_1 or the symmetric CO_3^{2-} stretching vibration [21]. Moreover, the two low-energy weak peaks at ~155 and 280 cm^{-1} included in the Raman spectrum of calcite arose from the external vibrations of the CO_3^{2-} groups that involved translatory and rotatory oscillations of those groups within the crystal lattice mode [22].

Gypsum and anhydrite were detected in the gesso ground layers of samples D3, D10, D12, D14, D15, and D17. The Raman spectra of gypsum showed a strong peak at 1008 cm^{-1}, which was attributed to the v1(a1) symmetric stretching vibration modes of SO_4^{2-} tetrahedra [14]. Anhydrite ($CaSO_4$,) is often found with gypsum, and a study of the

Raman spectrum of the former showed that the band attributed to the symmetric stretch vibration v1(a1) $SO_4{}^{2-}$ appeared at approximately 1016 cm^{-1} [14].

The Raman spectra of orpiment (As_2S_3) displayed the characteristic strong bands at 136, 154, 202, 293, 311, and 355 cm^{-1}. Each As atom was surrounded by three sulphur atoms, and each (S) atom was shared by two (As) atoms, while the layers were held together by van der Waals forces [23]. The region between 100 and 200 cm^{-1} were bands due to the S-As-S angle bending, while the one from approximately 300 to 400 cm^{-1} was the As–S stretching vibrations [24].

Figure 5 is shown as examples of the micro-Raman results. The identifications of pigments and their assignments in the sample's cross section D17 (Figure 2a,b), based on the photographs obtained using optical microscopy, are described in the figure caption. This type of analysis was carried out for all samples included in the study.

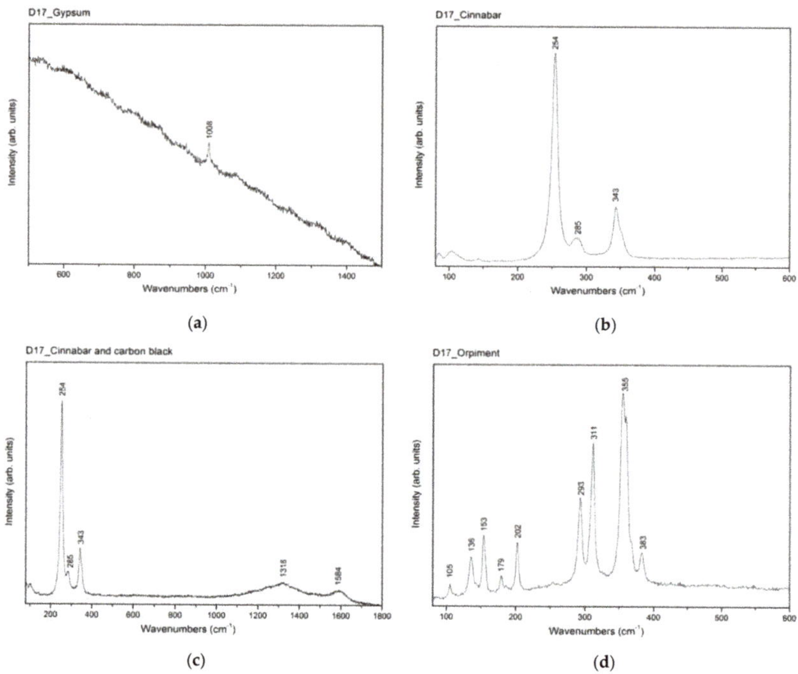

Figure 5. Four Raman spectra, which led to the identification of pigments and the gesso ground in the sample cross sections D17 (Figure 2a,b), are included as examples. In particular, the following were identified: (**a**) gypsum, (**b**) cinnabar, (**c**) carbon black with cinnabar, and (**d**) orpiment.

3.4. FTIR Results

Varnish samples were taken from two different spots of the door surface, sampling spots #13 and #16; their powder infrared spectra are shown in Figure 6.

In both samples, complex organic materials were detected, with natural resin (possibly mastic, a triterpenoid) being the most abundant common component, with strong maxima at 2956/2926 cm^{-1} and 2857 cm^{-1} (C-H stretching), 1738/1709 cm^{-1} (C=O stretch), 1458 cm^{-1} and 1384 cm^{-1} (CH_2, CH_3 bending vibrations), along with weaker ones at 1171, 1124, 1073, 1040, 952 cm^{-1} (generally, C-O stretch), and 744 cm^{-1} (C=C–H bend) [25–27]. Additionally, the relatively intense band at 848 cm^{-1} in sample D13 could also be assigned to aromatic C–H vibrations. The pronounced 2926 cm^{-1} component indicates additional material such as oil, or possibly another resin, while the feature at 1290–1280 cm^{-1} (stronger in sample D13) [28] could be assigned to aromatic ethers or phenolic moieties formed as a result of the heat shock. This, in combination with the weaker, typically aromatic bands at 3064 (w), 1608 (sh), 1513 (w) cm^{-1}, may account for the heat-induced aromatization of the triterpenoid resin [29,30] on the object's surface [31,32]. This may occur according to the following reaction scheme [30], showing the aromatization of ring A of oleanolic acid, a typical mastic component (Figure 7).

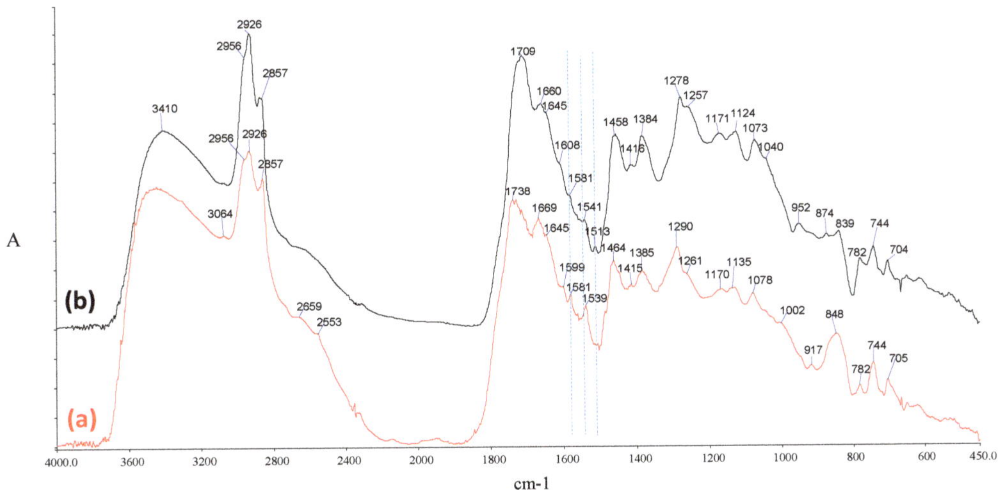

Figure 6. Infrared spectra of varnish powder samples; (**a**) D13 and (**b**) D16. Dashed vertical lines depict correspondence of aromatic and metal salt features between the two samples.

Figure 7. Possible oxidation route of oleanonic acid, a typical triterpenic mastic component leading to aromatic product (formation of aromatic ring A).

The weak but characteristic peaks observed at 1581 (sh), 1541 cm^{-1} (also stronger in sample D13) are indicative of the antisymmetric carboxylate vibration of metal salts [33–35], possibly due to interactions between Pb^{2+} and Ca^{2+} (due to lead white and gypsum, respectively, in the vicinity of the samples), and either the acidic fraction of the resin or a possible oil component, which, however, is not clearly detectable in the recorded spectra due to overlaps of its most prominent peaks (i.e., ester or acidic carbonyls and C-H vibrations) by those of the resin material. More specifically, the maxima can be assigned to the antisymmetric stretch (or v$_{as}$) of unidentate (~1580 cm^{-1}) and bidentate (~1540 cm^{-1}) calcium carboxylate complexes [33,36], while the weak peak at 1513 cm^{-1} (more prominent in sample #16), where lead white is detected, can be assigned to the v$_{as}$ of lead carboxylate [37,38].

The heat shock experienced by the varnish layer first led to the significant softening of the material, allowing for metal ion migration through the layers and the consequent formation of metal salts with the acid resin components.

4. Conclusions

Beyond its historical importance, including a documented fire incident [39] in 1535, the wood-carved door has a specific technological significance, which was investigated through this study. A striking feature of the door's decoration is the extended palette of inorganic and consecutive organic colourant layers, along with consecutive varnish layers; while in another sample from a nearby area, the alternating ground and pigment layers arguably suggested a correction at a later, unspecified time (samples D14 and D17).

The varnish samples showed a triterpenoid resin, possibly mastic, with added components (oil or other resin). Their infrared spectra showed signs of deterioration due to interactions of the acidic organic components, with the metal ions of pigments and the ground layer, and indications of heat-induced oxidations (aromatization); this may well be in accordance with the recorded history of the object involved in a fire incident.

As shown in Table 2, Raman spectra and SEM-EDX revealed the presence of a wide colourants palette, including yellow (orpiment), red (lead red, red ochre, and cinnabar), blue (indigo/woad), lead white, and carbon black; green could also be assigned to a copper pigment, although more evidence is needed. Furthermore, gypsum, anhydrite, and calcite were identified as the ground-layer materials in most samples.

As objects of the late Byzantine period have rarely been studied through multi-analytical physicochemical approaches [2–4], the data of Table 2 can be considered as a reference point for researchers who are interested in having a broad view of the Byzantine painting palette and techniques.

Author Contributions: A.K. conceptualization; methodology; investigation; data curation; writing original draft; I.K. supervision; resources; data curation; writing—review and editing; S.C.B. supervision; funding acquisition; data curation; writing—review and editing. All authors have read and agreed to the published version of the manuscript.

Funding: This work has been funded through the Scholarship for Doctoral Candidates by the Special Account For Research Grants, University of West Attica.

Institutional Review Board Statement: Not applicable.

Informed Consent Statement: Not applicable.

Acknowledgments: We would like to acknowledge the Abbot of the Monastery of St. Dionysiou in Mount Athos, Archimandrite Petros, and all the brotherhood. In addition, George Fousteris for his contribution in the historical field of a particular study, and Dimitrios Karakatsanis for providing us with valuable photos of the artifact. Finally, the authors would like to thank the Research Committee and the Special Account for Research Grants of the University of West Attica for financially supporting Konstanta's PhD studies.

Conflicts of Interest: The authors declare no conflict of interest.

References

1. Kadas, S.N. *Monastery of Saint Dionysios. Pilgrimage Guide (History-Art-Relics)*, 3rd ed.; Monastery of Saint Dionysios: Mount Athos, Greece, 2002; pp. 56–59.
2. Karapanagiotis, I.; Lampakis, D.; Konstanta, A.; Farmakalidis, H. Identification of colourants in icons of the Cretan School of iconography using Raman spectroscopy and liquid chromatography. *J. Archaeol. Sci.* **2013**, *40*, 1471–1478. [CrossRef]
3. Karapanagiotis, I.; Minopoulou, E.; Valianou, L.; Daniilia, S.; Chryssoulakis, Y. Investigation of the colourants used in icons of the Cretan School of iconography. *Anal. Chim. Acta* **2009**, *647*, 231–242. [CrossRef]
4. Valianou, L.; Wei, S.; Mubarak, M.S.; Farmakalidis, H.; Rosenberg, E.; Stassinopoulos, S.; Karapanagiotis, I. Identification of organic materials in icons of the Cretan School of iconography. *J. Archaeol. Sci.* **2011**, *38*, 246–254. [CrossRef]
5. Daniilia, S.; Andrikopoulos, K.S.; Sotiropoulou, S.; Karapanagiotis, I. Analytical study into El Greco's baptism of Christ: Clues to the genius of his palette. *Appl. Phys. A* **2008**, *90*, 565–575. [CrossRef]
6. Nusimovici, M.A.; Meskaou, A. Raman scattering by α-HgS (Cinnabar). *Phys. Stat. Sol. B* **1973**, *68*, 121–125. [CrossRef]
7. Bell, I.M.; Clark, R.J.H.; Gibbs, P.J. Raman spectroscopic library of natural and synthetic pigments (pre −1850 AD). *Spectrochim. Acta A* **1997**, *53*, 2159–2179. [CrossRef]
8. Mazzocchin, G.A.; Agnoli, F.; Salvadori, M. Analysis of Roman age wall paintings found in Pordenone, Trieste and Montegrotto. *Talanta* **2004**, *64*, 732–741. [CrossRef] [PubMed]

9. McCarty, K.F. Inelastic light scattering in a-Fe$_2$O$_3$: Phonon vs Magnon scattering. *Solid State Commun.* **1988**, *68*, 799–802. [CrossRef]

10. Vigouroux, J.P.; Husson, E.; Calvarin, G.; Dao, N.Q. Etude par spectroscopié vibrationnelle des oxydes Pb$_3$O$_4$, SnPb$_2$O$_4$ et SuPb(Pb$_2$O$_4$)$_2$. *Spectrochim. Acta Part A Mol. Spectrosc.* **1982**, *38*, 393–398. [CrossRef]

11. Tuinstra, F.; Koenig, J.L. Raman spectrum of graphite. *J. Chem. Phys.* **1970**, *53*, 1126–1130. [CrossRef]

12. Nakamizo, M.; Kammereck, R.; Walker, P.L., Jr. Laser Raman studies on carbons. *Carbon* **1974**, *12*, 259–267. [CrossRef]

13. Tatsch, E.; Schrader, B. Near-infrared Fourier Transform Raman spectroscopy of indigoids. *J. Raman Spectrosc.* **1995**, *26*, 467–473. [CrossRef]

14. Withnall, R.; Shadi, I.T.; Chowdhry, B.Z. Case study: The analysis of dyes by SERRS. In *Raman Spectroscopy in Archaeology and Art History*; Edwards, H.G.M., Chalmers, J.M., Barnett, N.W., Eds.; The Royal Society of Chemistry: Cambridge, UK, 2005; pp. 152–166.

15. Gunasekaran, S.; Anbalagan, G.; Pandi, S. Raman and infrared spectra of carbonates of calcite structure. *J. Raman Spectrosc.* **2006**, *37*, 892–899. [CrossRef]

16. Morimoto, N. The crystal structure of orpiment (As$_2$S$_3$) refined. *Mineral. J.* **1954**, *1*, 160–169. [CrossRef]

17. Forneris, R. The infrared and Raman spectra of realgar and orpiment. *Am. Mineral.* **1969**, *54*, 1062–1074.

18. Türker, M.C.Z.; Kısasöz, A.; Guler, K. Experimental Research on Properties of Naval Brass Castings. *Prac. Metallogr.* **2016**, *53*, 24–35. [CrossRef]

19. Sandu, I.; Murta, E.; Neves, E.; Pereira, M.; Sandu, A.V.; Kuckova, S.; Maurício, A. A comparative interdisciplinary study of gilding techniques and materials in two Portuguese Baroque "talha dourada" complexes. *Estud. Conserv. Restauro* **2013**, *1*, 47–71. [CrossRef]

20. Vermeulen, M.; Sanyova, J.; Janssens, K. Identification of artificial orpiment in the interior decorations of the Japanese tower in Laeken, Brussels, Belgium. *Herit. Sci.* **2015**, *3*, 9. [CrossRef]

21. Ogalde, J.P.; Salas, C.O.; Lara, N.; Leyton, P.; Paipa, C.; Vallette, M.C.; Arriaza, B. Multi-instrumental identification of orpiment in archaeological mortuary contexts. *J. Chil. Chem. Soc.* **2014**, *59*, 2571–2573. [CrossRef]

22. Rodríguez, S.L.; López, V.; Miguel, L.C. Microscopic Identification of Vine Black Pigment In A Tempera Painting By Francisco De Goya. *SM Anal. Bioanal. Tech.* **2017**, *2*, 1–8. [CrossRef]

23. Tanevska, V.; Nastova, I.; Minčeva-Šukarova, B.; Grupče, O.; Ozcatal, M.; Kavčić, M.; Jakovlevska-Spirovska, Z. Spectroscopic analysis of pigments and inks in manuscripts: II. Islamic illuminated manuscripts (16th–18th century). *Vib. Spectrosc.* **2014**, *73*, 127–137. [CrossRef]

24. Gómez, B.; Parera, S.; Siracusano, G.; Maier, M. Integrated analytical techniques for the characterization of painting materials in two South American polychrome sculptures. *E-Preserv. Sci.* **2010**, *7*, 1–7.

25. Theodorakopoulos, C. The Excimer Laser Ablation of Picture Varnishes. Ph.D. Thesis, Royal College of Art, London, UK, 27 May 2005.

26. Ménager, M.; Azémard, C.; Vieillescazes, C. Study of Egyptian mummification balms by FT-IR spectroscopy and GC-MS. *Microchem. J.* **2014**, *114*, 32–41. [CrossRef]

27. Nevin, A.; Comelli, D.; Osticioli, I.; Toniolo, L.; Valentini, G.; Cubeddu, R. Assessment of the ageing of triterpenoid paint varnishes using fluorescence, Raman and FTIR spectroscopy. *Anal. Bioanal. Chem.* **2009**, *39*, 2139–2149. [CrossRef]

28. Azémard, C.; Vieillescazes, C.; Ménager, M. Effect of photodegradation on the identification of natural varnishes by FT-IR spectroscopy. *Microchem. J.* **2014**, *112*, 137–149. [CrossRef]

29. Bruni, S.; Guglielmi, V. Identification of archaeological triterpenoid resins by the non-separative techniques FTIR and 13C NMR: The case of Pistacia resin (mastic) in comparison with frankincense. *Spectrochim. Acta Part A Mol. Biomol. Spectrosc.* **2014**, *121*, 613–622. [CrossRef]

30. Neilson, A.H.; Hynning, P. Polycyclic aromatic hydrocarbons: Products of chemical and biochemical transformation of Alicyclic precursors. *Toxicol. Environ. Chem.* **1996**, *53*, 45–89. [CrossRef]

31. Tirat, S. Propriétés Physico-Chimiques et Vieillissement des Vernis huile de lin/Colophane: De la Technique du Luthier à la Conservation des Instruments de Musique Vernis. Ph.D. Thesis, Universite De Cergy-Pontoise, Cergy-Pontoise, France, 8 December 2017.

32. Tirat, S.; Degano, I.; Echard, J.P.; Lattuati-Derieux, A.; Lluveras-Tenorio, A.; Arul, M.; Serfaty, S.; Huero, J.Y. Historical linseed oil/colophony varnishes formulations: Study of their molecular composition with micro-chemical chromatographic techniques. *Microchem. J.* **2016**, *126*, 200–213. [CrossRef]

33. Otero, V.; Sanches, D.; Montagner, C.; Vilarigues, M.; Carlyle, L.; Lopes, J.; Melo, M. Characterisation of metal carboxylates by Raman and infrared spectroscopy in works of art. *J. Raman Spectrosc.* **2014**, *45*, 11–12. [CrossRef]

34. Ma, X.; Beltran, V.; Ramer, G.; Pavlidis, G.; Parkinson, D.; Thoury, M.; Meldrum, T.; Centrone, A.; Berrie, B. Revealing the Distribution of Metal Carboxylates in Oil Paint from the Micro- to Nanoscale. *Angew. Chem.* **2019**, *131*, 11778–11782. [CrossRef]

35. Hermans, J.J. Metal Soaps in Oil Paint: Structure, Mechanisms and Dynamics. Ph.D. Thesis, Van 't Hoff Institute for Molecular Sciences (HIMS), University of Amsterdam, Amsterdam, The Netherlands, 9 May 2017.

36. Nakamoto, K. *Infrared and Raman Spectra of Inorganic and Coordination Compounds, Part B, Applications in Coordination, Organometallic, and Bioinorganic Chemistry*, 6th ed.; John Wiley & Sons: Hoboken, NJ, USA, 2009; pp. 64–65.

37. Plater, M.J.; De Silva, B.; Gelbrich, T.; Hursthouse, M.; Higgitt, C.; Saunders, D. The characterisation of lead fatty acid soaps in "protrusions" in aged traditional oil paint. *Polyhedron* **2003**, *22*, 3171–3179. [CrossRef]
38. Catalano, J.; Yao, Y.; Murphy, A.; Zumbulyadis, N.; Centeno, S.A.; Dybowski, C. Analysis of Lead Carboxylates and Lead-Containing Pigments in Oil Paintings by Solid- State Nuclear Magnetic Resonance. *MRS Online Proc. Libr.* **2014**, *1656*, 371–379. [CrossRef]
39. Kadas, S.N. *The Notes of the Manuscripts of the Monastery of Dionysios of Mount Athos*; Monastery of Saint Dionysios: Mount Athos, Greece, 1996; pp. 64–67.

Article

Compositional and Morphological Comparison among Three Coeval Violins Made by Giuseppe Guarneri "del Gesù" in 1734

Giacomo Fiocco [1,2], Sebastian Gonzalez [3], Claudia Invernizzi [1], Tommaso Rovetta [1], Michela Albano [1,4], Piercarlo Dondi [5], Maurizio Licchelli [1,6], Fabio Antonacci [3] and Marco Malagodi [1,7,*]

1   Arvedi Laboratory of Non-Invasive Diagnostics, CISRiC, University of Pavia, 26100 Cremona, Italy; giacomo.fiocco@unipv.it (G.F.); claudia.invernizzi@unipv.it (C.I.); tommaso.rovetta@unipv.it (T.R.); michela.albano@unipv.it (M.A.); maurizio.licchelli@unipv.it (M.L.)
2   Department of Chemistry, University of Torino, 10125 Torino, Italy
3   DEIB, Cremona Campus, Polytechnic of Milan, 20133 Milano, Italy; juansebastian.gonzalez@polimi.it (S.G.); fabio.antonacci@polimi.it (F.A.)
4   Department of Physics, Polytechnic of Milan, 20133 Milano, Italy
5   Department of Electrical, Computer and Biomedical Engineering, University of Pavia, 27100 Pavia, Italy; piercarlo.dondi@unipv.it
6   Department of Chemistry, University of Pavia, 27100 Pavia, Italy
7   Department of Musicology and Cultural Heritage, University of Pavia, 26100 Cremona, Italy
*   Correspondence: marco.malagodi@unipv.it; Tel.: +39-349-644-5217

Abstract: In the present work, we had the opportunity to study the coating systems of three different coeval violins, namely "Spagnoletti", "Stauffer", and "Principe Doria", made by Giuseppe Guarneri "del Gesù" in 1734. These three violins were non-invasively investigated by reflection Fourier transform infrared spectroscopy and X-ray fluorescence. These two techniques were combined for the first time with a 3D laser scanner. The analytical campaign enabled the characterization of the materials and their distribution within the stratigraphy, mainly composed of varnish and, when present, of a proteinaceous ground coat. Some restoration materials were also identified, suggesting the application of different maintenance treatments undertaken during their history. The preliminary information about morphological and geometrical differences between the three coeval violins were acquired through the 3D laser scanner in order to observe similarities and differences in the design features among the three violins.

Keywords: violin; XRF; FTIR; reflection infrared spectroscopy; 3D scan; stratigraphy; varnish; musical instrument

Citation: Fiocco, G.; Gonzalez, S.; Invernizzi, C.; Rovetta, T.; Albano, M.; Dondi, P.; Licchelli, M.; Antonacci, F.; Malagodi, M. Compositional and Morphological Comparison among Three Coeval Violins Made by Giuseppe Guarneri "del Gesù" in 1734. *Coatings* **2021**, *11*, 884. https://doi.org/10.3390/coatings11080884

Academic Editor: Nervo Marco

Received: 30 June 2021
Accepted: 19 July 2021
Published: 23 July 2021

Publisher's Note: MDPI stays neutral with regard to jurisdictional claims in published maps and institutional affiliations.

1. Introduction

In the 17th and 18th centuries, during the so-called Golden Age period of historical violin making, hundreds of musical instruments were made in the workshops of the most important luthiers based in Cremona (Italy), namely Antonio Stradivari, Nicola Amati, Giuseppe Guarneri "del Gesù", and other members of the great luthier families. Traditional manufacturing techniques, materials, and recipes employed in violin making were passed down from master artisans to apprentices, gradually changing through the decades. The decline of the art of violin making in Cremona, which occurred from the beginning of the 19th century, caused the loss of these traditions but left us numerous masterpieces of inimitable beauty. Nevertheless, starting from the 20th century, the high interest in the topic has led some violin makers and scientists to study the lost manufacturing techniques and the materials used by the old Cremonese Masters [1,2]. In recent years, some research teams have considered the investigation with a scientific approach [3–5] with the aims of characterizing the nature of materials involved in the Cremonese finishing treatments, i.e., the varnish [6,7], and understanding how they affect the acoustical properties of the musical instrument [8–10]. For this purpose, a large set of non-invasive and micro-invasive techniques—when possible—were widely used to investigate the physical and chemical features of the materials. The most common techniques were X-ray fluorescence (XRF) [11,12], Fourier transform infrared (FTIR) spectroscopy in different geometries [13–17], tomographic techniques, like computed axial tomography (CAT) or optical coherence tomography (OCT), imaging techniques, and chromatographic and different spectroscopic analyses [18–20].

Under this scenario, analytical investigations enabled researchers to obtain a clear picture of the expected materials and their multi-layered coating systems, consisting in multiple superimposed films of varnishes, generally applied on wood, which was previously treated with a sealer to prevent varnish penetration. The most common materials involved in the finishing processes are siccative oils, metal-based driers, natural resins, animal glues, inorganic fillers, organic colorants, inorganic pigments, and lake pigments [21,22].

In this research, we had the opportunity to study and compare three esteemed coeval violins, namely the "Spagnoletti", the "Stauffer", and the "Principe Doria" (Figure 1 and Figure S1) made by Giuseppe Guarneri "del Gesù" (Cremona, 1698—1744) in 1734. They were made in the period between 1730 and 1742, the most fruitful for the Master [23]. Instruments of this period are considered by contemporary violin makers unsurpassed specimens of established shape, with faultless workmanship [24].

Figure 1. Visible (left) and UV (right) images of the soundboard of (**a**) the "Spagnoletti", (**b**) the "Stauffer", and (**c**) the "Principe Doria" made by Giuseppe Guarneri "del Gesù" in 1734. Visible and UV images of the back plates are shown in Figure S1 in the Supplementary Materials.

A spectroscopic non-invasive approach through reflection FTIR and XRF was combined with a 3D model obtained with a 3D laser scanner. XRF and FTIR allowed us to identify the elemental and molecular composition of the materials, and the 3D scan provided decisive information regarding the constructive features such as arching, corners, and f-hole shapes. To our knowledge, an analytical strategy that combines a diagnostic campaign focused on material characterization with a structural assessment of three contemporary instruments with such a high historical significance has never been accomplished before.

The present study intends to (i) non-invasively characterize the chemical composition of the materials as well as understand their distribution within the coating systems, (ii) obtain preliminary valuable information about the design peculiarities adopted by Guarneri del Gesù in 1734 when he built the "Spagnoletti", the "Stauffer", and the "Principe Doria" violins, and, finally, (iii) observe similarities and differences in the material and design features among the three violins.

2. Experimental

Visible and ultraviolet-induced fluorescence (UVIFL) images were acquired following the protocol described in [25]. Visible illumination was provided by two soft-box LEDs (T = 5400K), while UV illumination was provided by two Wood's lamp tubes TL-D 36 W BBL IPP low-pressure Hg tubes, 40 Watt, emission peak approximately 365nm (Philips, Amsterdam, Netherlands). Photos were taken with a Nikon D4 full-frame digital camera (Nikon Corporation, Tokyo, Japan) with a 50mm f/1.4 Nikkor objective (Nikon Corporation, Tokyo, Japan) (for visible: 1/10-1/15 exposure time, aperture f/8, ISO 100; for UV: 30s exposure time, aperture f/8, ISO 400). Violins were vertically placed (using the end-button hole) on a laboratory-built rotating platform, moved by a stepper motor (NEMA 17, Changzhou Jinsanshi Mechatronics Co. Ltd., Changzhou, China) controlled with an EasyDriver V44 driver board, in turn controlled by an Arduino Nano board. A simplified graphical interface was also developed to send instruction to the Arduino in a comfortable way [25].

For the preliminary analysis of the UVIFL images, we used UV Analyzer, an interactive tool developed in our lab that implements a series of image processing algorithms able to quickly highlight and label regions of interest on the surface of the instrument [25,26].

Reflection FTIR spectroscopy was performed using the Alpha portable spectrometer (Bruker Optics, Billerica, MA, USA) equipped with the R-Alpha external reflectance module that is composed of an optical layout of $23°/23°$. The compact optical bench was composed of a SiC Globar source (Bruker Optics, Billerica, MA, USA), a RockSolid interferometer (with gold mirrors, Bruker Optics, Billerica, MA, USA), and an uncooled DLaTGS detector (Bruker Optics, Billerica, MA, USA). The reflection module allowed contactless and non-invasive analysis of measurement areas of about 5 mm in diameter at a working distance of 15 mm. Pseudo-absorbance spectra [log(1/R); R—reflectance] were collected between 7500 and 375 cm^{-1}, at a resolution of 4 cm^{-1} and with an acquisition time of 1 min. The background was acquired using a gold flat mirror. For the study of derivative bands, reflection infrared spectra were transformed to absorbance spectra by applying the Kramers–Kronig transformation (KKT) included in the OPUS software package (7.2 version, 2013, Bruker, Billerica, MA, USA) and a portion of the mid-IR spectral range (3600–400 cm^{-1}) is considered.

X-Ray fluorescence (XRF) spectroscopy was performed in single-point mode, using an ELIO portable X-ray fluorescence spectrometer by Bruker Optics Corporation (Billerica, MA, USA). It uses an X-ray source working with an Rh anode and an analytical spot diameter of 1.3 mm. Spectra were collected on 2048 channels at 40 kV, 80 µA, 480 s. For each element taken into consideration, the displayed value corresponds to the net area counts of the peak (Kα) normalized to time and to the average of net area counts of the coherent scattering Rh-Kα peak [27,28]. Data processing was carried out with ELIO software (1.6.0.57 version, 2020, Bruker, Billerica, MA, USA).

The 3D scan of the violins was performed using a RS3 Integrated Scanner (a linear laser scanner with a stated accuracy of 30 µm) mounted on a mobile arm with 7 degrees of freedom (Romer Absolute Arm 7-Axis "SI"), both produced by Hexagon Metrology (Cobham, United Kingdom). PolyWorks suite (12.1, 2012, InnovMetric, Quebec City, Canada) was used as the main scanning and editing software. Blender (2.79, 2017, Blender, Amsterdam, Netherlands) was then used for some final refinements of the meshes. Before the 3D scan, all the mobile components of the violin that could interfere with the acquisition (the strings, the tailpiece, the bridge, the chin rest, and the pegs) were removed. To maintain a stable position of the violins during the scanning process without damaging the varnishes, we used two Plexiglas supports provided with clamps covered with gum. During the acquisition, the laboratory was maintained in a stable climatic condition with a temperature of 20 °C and a humidity of 50% (the same as the showcases of the museum). All the steps of the acquisition and editing procedure were described in [29]. For the analysis of the scans, we used the 3D point cloud and post-processed them with Mathematica (11.3, 2019, Wolfram Research, Champaign, IL, USA). To smooth the data of the arching (discussed in Section 3.2), we subsampled the data using half the available points. As for the corners (discussed in Section 3.2), we computed the alpha shape with a resolution of 5 mm. The outline plots (discussed in Section 3.2) are not smoothed, as the variations are smaller than the line thickness at the scale we plotted. All the tools used are discussed in more detail in [30,31].

3. Results and Discussion

3.1. Non-Invasive Analytical Investigation by Reflection FTIR and XRF

The UVIFL images allowed highlighting the best-conserved (BC), worn-out (WO), and restored (RE) areas [11], according to their different UV fluorescence levels (Figure 1 and Figure S1). These images were also run through the UV Analyzer [25] (Figure 2 and Figure S2) to precisely select the analytical spots. Three areas were identified on the soundboard and five on the back plate of the "Spagnoletti"; three on the soundboard and five on the back plate of the "Stauffer"; five on the soundboard and four on the back plate of the "Principe Doria". In order to obtain a complete stratigraphic picture of the different materials spreading from the external layers to the deeper ones, all the BC, WO, and RE areas were analyzed ("Spagnoletti": 35 FTIR spots, 20 XRF spots; "Stauffer": 35 FTIR spots, 16 XRF spots; "Principe Doria": 31 FTIR spots, 15 XRF spots).

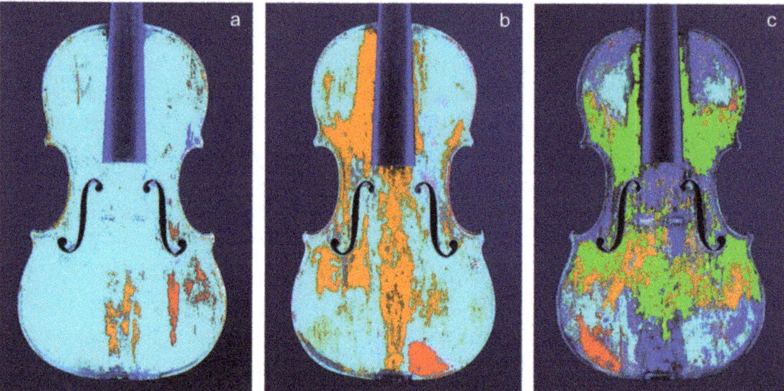

Figure 2. Images of the soundboard of (**a**) the "Spagnoletti", (**b**) the "Stauffer", and (**c**) the "Principe Doria" violins obtained through UV Analyzer. Different colors correspond to best-conserved (orange), worn-out (light blue), and restored (red, blue, and green) areas, according to their different UV-induced fluorescence colors. Areas identified on the back plates are shown in Figure S2 in the Supplementary Materials.

Reflection FTIR spectra collected on WO areas of the "Principe Doria" and the "Stauffer" show characteristic signals of a proteinaceous material, possibly casein or animal glue [18], which had been spread on the entire body of the violins before varnishing. As regards the "Spagnoletti", instead, proteins were only revealed in circumscribed areas of the upper and lower corners, suggesting their correlation with subsequent restoration work. The KKT diagnostic bands of this compound fall at 3310 cm^{-1} (ν_{as}NH), 1660 cm^{-1} (νC=O or amide I), and 1550–45 cm^{-1} (combination of νCN and δNH or amide II) [32–35] (Figure 3a).

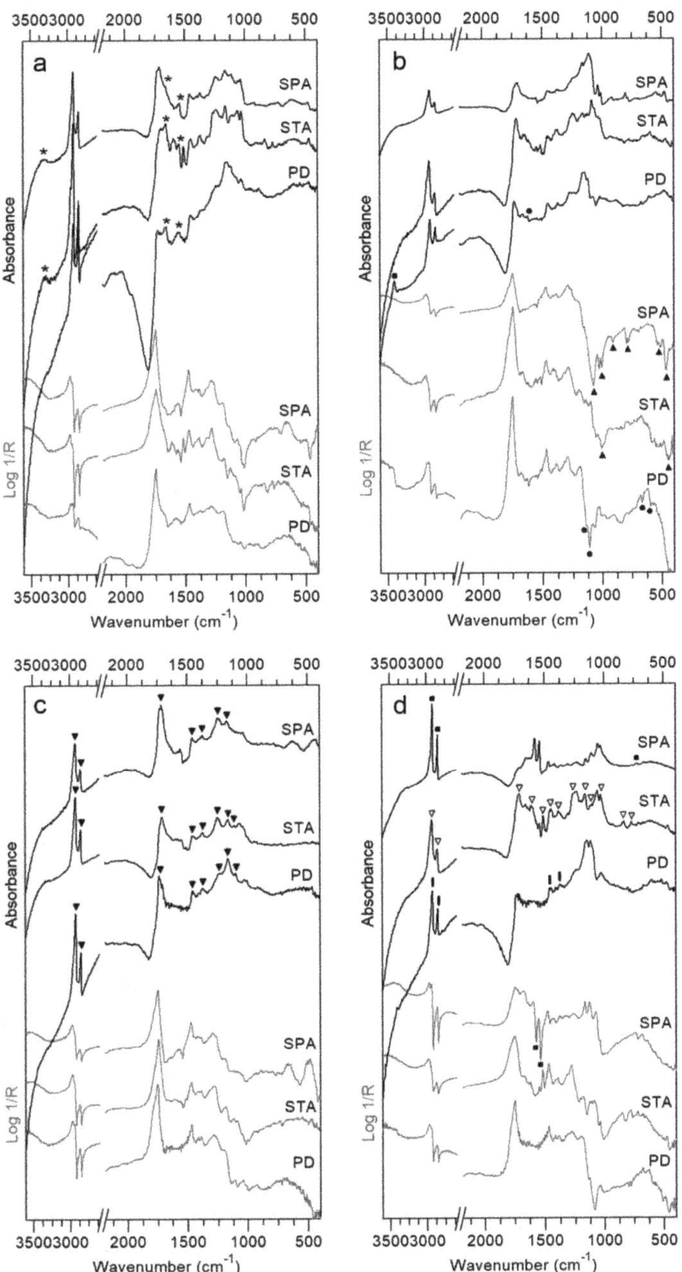

Figure 3. Reflection FTIR spectra in pseudo-absorbance (gray line) and after KKT (black line) collected in areas of the three violins (SPA—"Spagnoletti", STA—"Stauffer", PD—"Principe Doria") containing: (**a**) proteins, (**b**) inorganic compounds, (**c**) varnishes, and (**d**) restoration substances. Marker bands of proteins (∗), silicates (▲), sulfates (●), varnishes (▼), metal soap (■), benzoin resin (▽), and aliphatic hydrocarbon (׀) are highlighted.

Some inorganic compounds were also investigated in the pseudo-absorbance spectra of the wide FTIR data set. In particular, silicate-based materials were detected in WO areas of the "Spagnoletti": bands at 3625 cm^{-1} (derivative, νOH inner hydroxyl groups), 1030 and 1005 cm^{-1} (inverted, ν_{as}SiO), 910 cm^{-1} (inverted, δOH inner hydroxyl groups), and 535 and 465 cm^{-1} (inverted, δAl–O–Si) are attributed to kaolin [14], whereas signals at 1080 cm^{-1} (inverted, ν_{as}SiO), 800 and 780 cm^{-1} (ν_sSiO), and ca. 450 cm^{-1} (inverted, δSiO) derive from quartz [14] (Figure 3b). Concerning the "Principe Doria", features of quartz (not shown here) were identified together with those of sulfate compounds, likely gypsum, with bands being positioned at 3410 cm^{-1} (derivative, νOH), 1140 and 1115 cm^{-1} (inverted, ν_{as}SO$_4$), and 670 and 600 cm^{-1} (inverted δ_{as}SO$_4$) [14] (Figure 3b). As already highlighted in [14,18], however, we cannot assign the sulfates to a precise layer of the "Principe Doria" stratigraphy due to their heterogeneous results on the plates. Conversely, FTIR spectra collected on the "Stauffer" did not enable a reliable identification of inorganic compounds, and only a tentative attribution to silicate-based minerals was possible (Figure 3b). XRF counts of low-weight elements Si and S, even if underestimated (atomic number Z < K), increased in the WO areas. These results can further confirm the presence of silicate- and sulfate-based phases dispersed in the ground coat or filling the wood pores.

Within each violin, the BC areas of the soundboard and back plate exhibit similar FTIR spectral profiles. In detail, the spectra collected on the "Principe Doria" display the KKT characteristic bands of an oil–resinous varnish at 2955 cm^{-1} (shoulder, ν_{as}CH$_3$), 2925 cm^{-1} (ν_{as}CH$_2$), 2855 cm^{-1} (ν_sCH$_2$), 1730–10 cm^{-1} (νC=O), 1460 cm^{-1} (δCH$_2$), 1380 cm^{-1} (δCH$_3$), 1250–40 and 1170–65 cm^{-1}, and 1100 cm^{-1} (νC–O) [14] (Figure 3c), possibly laid on the ground. If these signals can be definitely attributed to a mixture of oil(s) and vegetal resin(s) in the "Principe Doria", the identification is more uncertain in the "Spagnoletti" and the "Stauffer" due to the strong spectral similarities between oil-based varnish and shellac resin [32]. Interestingly, XRF analysis performed on the same BC areas highlighted an increase in counts of Fe and Mn and a possibly a correlation between the two elements in the "Spagnoletti" and the "Stauffer", suggesting the presence of iron-based pigments like umber earth [11] in the varnish. Conversely, these signals underwent only slight and non-significant variations between WO and BC areas of the "Principe Doria", probably due to invasive restoration processes, which removed the upper part of the external pigmented varnish [18]. In addition, higher concentrations of Pb were detected in some spots on the soundboard and back plate of the three violins, suggesting the use of a lead-based siccative dispersed in the oil binder [12].

As regards the restored areas, reflection FTIR highlighted KKT signals of shellac (not shown here), mainly in the RE areas of the "Spagnoletti" and the "Stauffer", on both sides of the body. The characteristic bands of this animal resin fall at 2925 (ν_{as}CH$_2$), 2855 cm^{-1} (ν_sCH$_2$), 1730 cm^{-1} (νC=O esters), 1710 cm^{-1} (νC=O acids), 1460 cm^{-1} (δCH$_2$), 1375 cm^{-1} (δCH$_3$), 1255 cm^{-1} (νC–O esters), 1155 cm^{-1} (νC–O acids), 940 and 925 cm^{-1} (δH$_2$C=C), and 720 cm^{-1} (δ_{as}CH$_2$) [15,32,36]. Additionally, a metal soap identifiable as calcium stearate was found in some RE and WO areas on both the soundboard and back plate of the "Spagnoletti" violin, and residues of benzoin resin—commonly used to polish a varnished surface—were revealed in the same areas of the "Stauffer". The detected marker bands of the calcium soap, acting as a lubricant, are located in the pseudo-absorbance spectra at 2955 cm^{-1} (derivative, ν_{as}CH$_3$), 2920 cm^{-1} (derivative, ν_{as}CH$_2$), 2850 cm^{-1} (derivative, ν_sCH$_2$), 1575 cm^{-1} (inverted, ν_{as}COO–), 1540 cm^{-1} (inverted, ν_sCOO–), and 720 cm^{-1} (derivative, δCH$_2$) [37,38] (Figure 3d). As reported in [37,38], the strong doublet at 1575 and 1540 cm^{-1} corresponds to the asymmetric carboxylate stretching vibrations of the calcium salts. The KKT diagnostic bands of benzoin resin were identified at 2930 and ca. 2865 cm^{-1} (νCH), 1710 cm^{-1} (νC=O conjugate ketone), 1610, 1600, and 1515 cm^{-1} (aromatic skeletal), 1460 and 1450 cm^{-1} (aromatic skeletal + δCH), ca. 1380 cm^{-1} (δCH), 1265, 1165, 1120, and 1030 cm^{-1} (νC–O), and 835, 770, and 710 cm^{-1} (undefined) [18] (Figure 3d). As regards the "Principe Doria", few FTIR analytical areas highlighted markers of benzoin resins and shellac. Interestingly, an aliphatic hydrocarbon compound (e.g., paraffin) was further singled out (Figure 3d). In the absence of some markers of the varnish (mainly $vC=O$ and $vC-O$), or when their intensity is low, the very sharp and intense CH_2 peaks can be attributed to long aliphatic chains [39,40]. Furthermore, evidence of sulfate-based phases, such as gypsum, was identified in heterogeneously diffused areas of the instrument. The presence of the latter inorganic compound was difficult to locate in a precise layer since it occurs in some areas at different varnish thicknesses. XRF signals of Br detected on the three violins can be attributed to bromomethane, commonly used in the recent past as a treatment against woodworms [41]. All these different materials highlighted on the surfaces of the three violins, commonly used for maintenance practice and as surface polish in restoration, are significant signs of different maintenance treatments undertaken during their long history.

Elemental composition of purflings was studied by XRF and the results obtained are reported in Figure 4. Measurements highlight the presence in all the instruments of significant amounts of Fe, Cu, and Zn, typically associated with iron-based dyes/stain [42]. Other elements, namely Si, S, K, Ca, Ti, Mn, Pb, and Ba, were also identified in the same analytical spots but they are not considered in the discussion since they are not related to the dyeing process but probably to the wood and/or other compounds, such as fillers, used in the finishing treatments. Depending on the composition of the raw materials and on the procedures used to mix them, the ratios between the elemental abundances can vary. In any event, purfling batches made in the same period, and thus stained with the same methods and substances, should generally show a comparable amount of marker elements (Fe, Cu, and Zn).

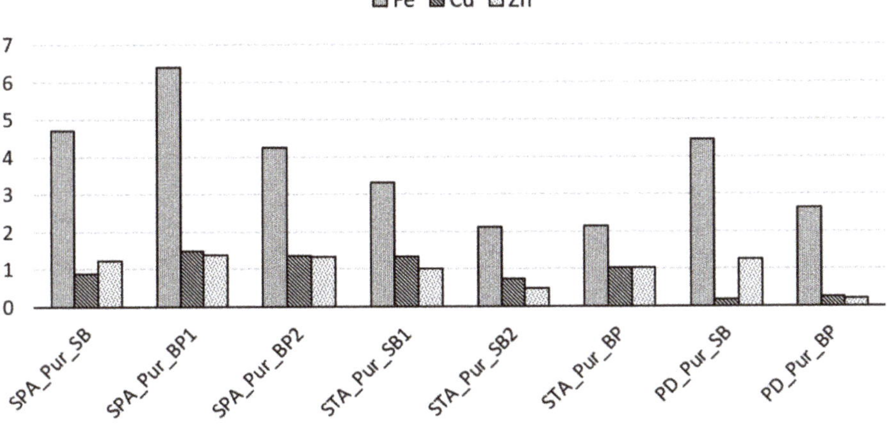

Figure 4. Estimates of Fe, Cu, and Zn detected by XRF analysis, corresponding to the purflings of the soundboard (SB) and back plate (BP) of the "Spagnoletti" (SPA), the "Stauffer" (STA), and the "Principe Doria" (PD). The y-axis values correspond to the net area counts of the element peaks ($K\alpha$) normalized to time and net area counts of the Rh peak ($K\alpha$).

The results obtained on the soundboard (SPA_Pur_SB in Figure 4) and back plate (SPA_Pur_BP1, BP2 in Figure 4) of the "Spagnoletti" and the "Stauffer" (STA_Pur_SB1, SB2, BP in Figure 4) purflings highlight, respectively, a similar amount of the selected marker elements (Figure 4). This could be attributable to a similar wood-dyeing procedure and, thus, probably to the same batch. On the other hand, if we compare the results of the two violins, slight variations of net area counts can be observed in the Fe results, as well as in the results obtained on the "Principe Doria". In fact, purflings of the "Principe Doria" revealed higher counts of Fe and lower counts of Cu, possibly due to a different iron-based dyeing process preparation procedure.

3.2. Morphological Investigation through 3D Laser Scanner

The outlines of the three violins (Figure 5a) were computed using the scan of the rib sections, as they provide a more accurate representation since they see less wear compared to the top or the back. To center and align the violins, we computed the moment of inertia of the point cloud, centered the violin with respect to the center of mass of the 3D scan, and rotated them along their principal axis. The computation was done with the whole violin, so the origin of the y-axis was in the upper part of the violin body rather than in the center of the violin. The measures of the three violins are shown in Table S1 in the Supplementary Materials.

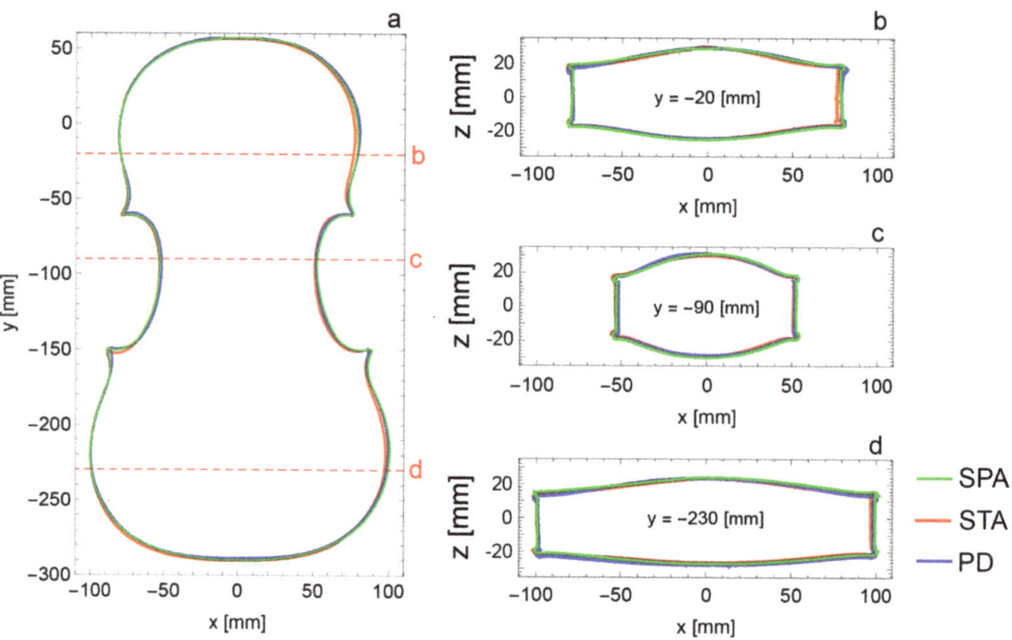

Figure 5. Outlines (**a**) and cross sections at different y locations (**b–d**) of the "Spagnoletti" (SPA, green line), the "Stauffer" (STA, red line), and the "Principe Doria" (PD, blue line).

The outlines of the violins studied here are quite similar, yet the lower left corner presents the largest variation between them. The measures collected corresponding to the cross sections shown in Figure 5b–d revealed some slight variations between the violins in the upper and lower right parts (Figure 5b,d; Figure S3f,h). As reported in Table S1, the "Principe Doria" revealed the maximum width in both measurements, while the "Stauffer" and the "Spagnoletti" are smaller. In particular, the "Stauffer" shows a significant variation both in the upper and lower parts of the soundboard, while the "Spagnoletti" only in the upper part. The width at the center (Figure S3g; Table S1g) is similar for all three. A different picture appears when looking at the back plate (Figure 5; Figure S3f,g,h). Here, the "Spagnoletti" shows the maximum values in the three measures while the "Principe Doria" and the "Stauffer" revealed slightly smaller values. As for the total lengths of the back plates of the "Spagnoletti", the "Stauffer", and the "Principe Doria", they are, respectively, 349.5 mm, 349.37 mm, and 349.95 mm, and those of the soundboard are 348.82 mm, 349.38 mm, and 348.81 mm. Despite these slight differences, the creation of the 3D models reveals a good match between the shape of the bodies, suggesting the use of the same mold and templates for all the three violins.

Figure 5b–d show cross sections of the violins at three different y locations: −20 mm, −90 mm, and −230 mm. The violins have roughly the same arching profile, except for the central part of the "Principe Doria", which exhibits a higher curvature on the left side. This is where the bass bar is located, and it could be due to a poorly fitted bass bar at some point in the history of the violin or to a conscious choice by the maker. The ribs present some places where they are not perfectly vertical, which is most visible at y = −20 mm (Figure 5b, green and blue lines) on both sides of the "Spagnoletti" and the "Principe Doria" and at y = −230 mm on the left side of the "Principe Doria" (Figure 5d, blue line). As regards the rib height, the three violins exhibit a consistent value of around 30 mm (Table S1i,j,j1,k), with that of the "Spagnoletti" seeming to be slightly higher than others. The small variations in the relative height are likely due to twisting of the wood and to the invasive restoration interventions that the instruments have undergone over the decades.

The corners of the top plates show a distinct wear pattern on the right side of the instrument, both in the upper and lower ones (Figure 6, solid lines). This wear pattern is most visible in the right side of the "Spagnoletti" (green solid line), where the bottom part of the upper corner has been worn till completely losing its shape. The lower corners look more dissimilar, in particular in the case of the "Stauffer", where the variation in the outline is also more noticeable. This aspect could be related to the high concert activity that the violins have undergone, whose effects of wear mainly fall on that side of the violin.

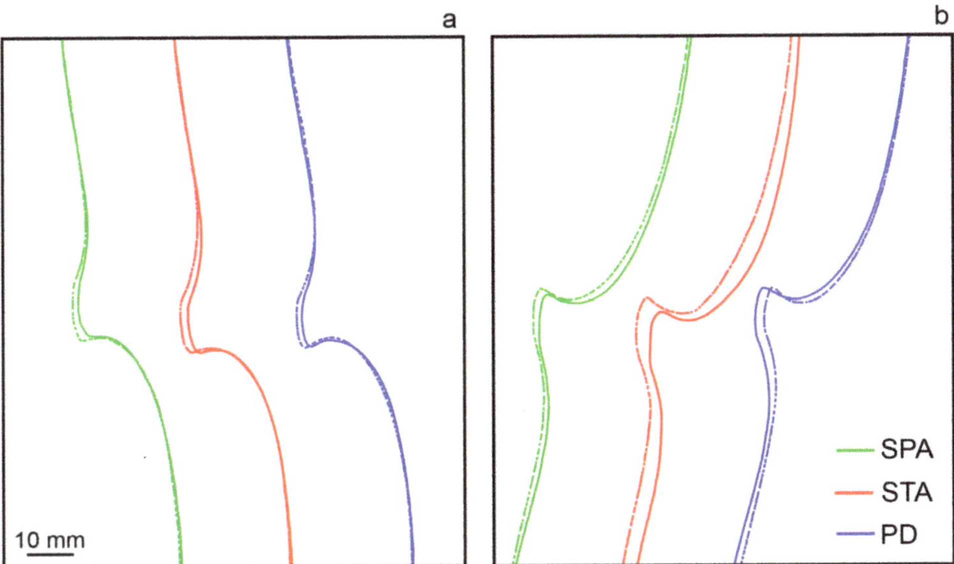

Figure 6. Outlines of the upper (**a**) and lower (**b**) corners of the "Spagnoletti" (SPA), the "Stauffer" (STA), and the "Principe Doria" (PD). In both images, the dotted lines are related to the left corners while the solid lines to the right ones.

The f-holes are the most different features among the three violins (Figure 7) in terms of position and dimensions. As reported in Table S1, the f-holes of the "Spagnoletti" and the "Stauffer" show similar a length of around 77 mm, whereas the ones of the "Principe Doria" are noticeably smaller in the total length (right hole: 74.59 mm, left hole: 75.41 mm). In terms of their width at the center (o, p, q in Table S1), they present similar values, except for the lower lobe distance in the "Principe Doria", where a larger variation between left and right holes is present.

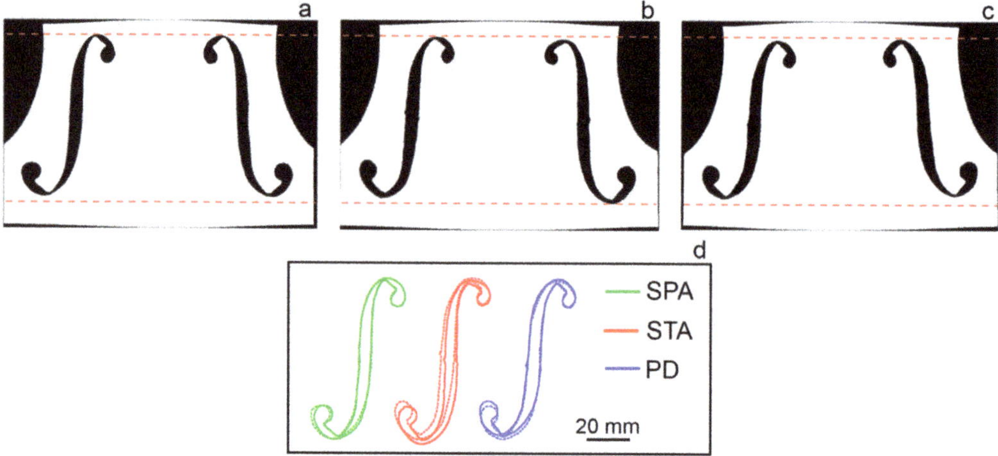

Figure 7. F-hole shapes of the "Spagnoletti" (**a**), the "Stauffer" (**b**), and the "Principe Doria" (**c**) and their position on the soundboard between the waists. Red dotted lines mark the maximum upper and lower position of the f-holes. Outlines of the f-holes (**d**) of the "Spagnoletti" (SPA), the "Stauffer" (STA), and the "Principe Doria" (PD).

In Figure 7d, the f-holes are mirrored and compared for each violin. The f-holes of the "Stauffer" and the "Principe Doria" are less symmetric, with either the left or the right one slightly higher than the other and, in the case of the "Stauffer", slightly out of alignment (left hole higher, right hole lower).

4. Conclusions

The combination of non-invasive spectroscopic approaches, like reflection FTIR and XRF, with the 3D laser scanner, allowed us to investigate both chemical and morphological features of three historical violins. In fact, the three masterpieces made by Giuseppe Guarneri del Gesù in 1734, namely the "Spagnoletti", the "Stauffer", and the "Principe Doria", were deeply characterized in terms of the finishing treatments, such as varnishes, ground coats, inorganic fillers, dispersed pigments, and restoration materials, as well as in the morphological and geometrical features, such as the shape of the body, the arching, the corners, and the f-holes. These are decisive peculiarities that help to better understand the artisanship of Guarneri "del Gesù".

Integrating and comparing the results, it is possible to observe how the three coeval violins underwent numerous restorations aimed to maintain the aesthetics of the finishing treatment. These repeated interventions led to the almost total removal or coverage of the original materials on most of the investigated areas. On the contrary, the geometry of the body remained almost unchanged, except for the central part of the "Principe Doria", which exhibits a higher curvature where the bass bar is located.

Lastly, the innovative combination of non-invasive spectroscopic techniques and the 3D laser scanner has proved to be a powerful tool to deeply characterize both the compositional and morphological features, providing detailed information about the finishing layers and past restoration actions, besides disclosing specific aspect of the shapes, morphology, and geometries of the violins. Moreover, the unique chance to directly compare the results of three instruments by the same maker and from the same year of production allowed us to make a step forward in the understanding of the Guarneri manufacturing, which seems to share common features in different instruments. Indeed, the outcome regarding the Master's manufacturing also plays a key role for modern violin makers, who often attempt to reproduce the excellent features of the Cremonese historical lutherie.

Supplementary Materials: The following are available online at https://www.mdpi.com/article/10.3390/coatings11080884/s1, Figure S1: Visible (left) and UV (right) images of the backplate of the Spagnoletti (a), Stauffer (b), and Principe Doria (c) made by Giuseppe Guarneri del Gesù in 1734. Figure S2: Images of the soundboard of the Spagnoletti (a), Stauffer (b), and Principe Doria (c) violins obtained through UV Analyzer. Different colors correspond to best-conserved (orange), worn-out (light blue), and restoration (red, blue, and green) areas, according to their different UV fluorescence levels. Figure S3: Legend of the measurement positions acquired on the violins. Table S1: Measurements obtained on the 3D models of the three violins. Measurement letters in the first column are related to the measurement positions shown in Figure S3.

Author Contributions: Conceptualization, G.F. and M.M.; methodology, G.F. and M.M.; software, G.F., C.I., T.R., M.A., P.D. and S.G.; validation, G.F., C.I., T.R., M.A., P.D., S.G., F.A., M.L. and M.M.; formal analysis, G.F., C.I., T.R., M.A. and P.D.; investigation, G.F., C.I. and T.R.; resources, M.M., F.A., and M.L.; data curation, G.F., C.I. and T.R.; writing—original draft preparation, G.F., C.I. and S.G.; writing—review and editing, G.F., C.I., T.R., M.A., P.D., S.G., F.A., M.L. and M.M.; visualization, G.F. and C.I.; supervision, M.M. and F.A. All authors have read and agreed to the published version of the manuscript.

Funding: This research received no external funding

Institutional Review Board Statement: Not applicable.

Informed Consent Statement: Not applicable.

Data Availability Statement: Data are contained within the article or Supplementary Materials.

Acknowledgments: We would like to thank the Fondazione Museo del Violino Antonio Stradivari, Cremona, for its collaboration and availability, which contributed to improving this research study. We are grateful to the "Friends of Stradivari" partners and to the owners of the violins for their important collaboration in this study.

Conflicts of Interest: The authors declare no conflict of interest.

References

1. Sacconi, S.F. *I segreti di Stradivari. Con il catalogo dei cimeli stradivariani del Museo civico "Ala Ponzone" di Cremona*; Libreria del Convegno: Cremona, Italy, 1979.
2. Michelman, J. *Violin Varnish: A Plausible Re-Creation of the Varnish Used by the Italian Violin Makers Between the Years 1550 and 1750, A.D.*, 1st ed.; Joseph Michelman: Cincinnati, OH, USA, 1946; pp. 58–154.
3. Tai, B.H. Stradivari's Varnish: A review of scientific findings—Part I. *J. Violin Soc. Am.* **2007**, *21*, 119–144.
4. Tai, B.H. Stradivari's Varnish: A review of scientific findings—Part II. *J. Violin Soc. Am.* **2009**, *22*, 1–31.
5. Von Bohlen, A.; Meyer, F. Microanalysis of old violin varnishes by total-reflection X-ray fluorescence. *Spectrochim. Acta Part B At. Spectrosc.* **1997**, *52*, 1053–1056. [CrossRef]
6. Echard, J.-P.; Bertrand, L.; Von Bohlen, A.; Le Hô, A.-S.; Paris, C.; Bellot-Gurlet, L.; Soulier, B.; Lattuati-Derieux, A.; Thao, S.; Robinet, L.; et al. The Nature of the Extraordinary Finish of Stradivari's Instruments. *Angew. Chem. Int. Ed.* **2009**, *49*, 197–201. [CrossRef]
7. Nagyvary, J.; Ehrman, J.M. The composite nature of the antique Italian varnish. *Naturwissenschaften* **1988**, *75*, 513–515. [CrossRef]
8. Lämmlein, S.; Künniger, T.; Rüggeberg, M.; Schwarze, F.W.; Mannes, D.; Burgert, I. Frequency dependent mechanical properties of violin varnishes and their impact on vibro-mechanical tonewood properties. *Results Mater.* **2021**, *9*, 100137. [CrossRef]
9. Lämmlein, S.L.; Mannes, D.; Van Damme, B.; Burgert, I.; Schwarze, F.W. Influence of varnishing on the vi-bro-mechanical properties of wood used for violins. *J. Mater. Sci.* **2019**, *54*, 8063–8095. [CrossRef]
10. Setragno, F.; Zanoni, M.; Antonacci, F.; Sarti, A.; Malagodi, M.; Rovetta, T.; Invernizzi, C. Feature-Based Analysis of the Impact of Ground Coat and Varnish on Violin Tone Qualities. *Acta Acust. United Acust.* **2017**, *103*, 80–93. [CrossRef]
11. Rovetta, T.; Invernizzi, C.; Licchelli, M.; Cacciatori, F.; Malagodi, M. The elemental composition of Stradivari's musical instruments: New results through non-invasive EDXRF analysis. *X-Ray Spectrom.* **2018**, *47*, 159–170. [CrossRef]
12. Echard, J.-P. In situ multi-element analyses by energy-dispersive X-ray fluorescence on varnishes of historical violins. *Spectrochim. Acta Part B At. Spectrosc.* **2004**, *59*, 1663–1667. [CrossRef]
13. Fiocco, G.; Invernizzi, C.; Grassi, S.; Davit, P.; Albano, M.; Rovetta, T.; Stani, C.; Vaccari, L.; Malagodi, M.; Licchelli, M.; et al. Reflection FTIR spectroscopy for the study of historical bowed string instruments: Invasive and non-invasive approaches. *Spectrochim. Acta Part A Mol. Biomol. Spectrosc.* **2021**, *245*, 118926. [CrossRef]
14. Invernizzi, C.; Fichera, G.V.; Licchelli, M.; Malagodi, M. A non-invasive stratigraphic study by reflection FT-IR spectroscopy and UV-induced fluorescence technique: The case of historical violins. *Microchem. J.* **2018**, *138*, 273–281. [CrossRef]

15. Invernizzi, C.; Daveri, A.; Vagnini, M.; Malagodi, M. Non-invasive identification of organic materials in historical stringed musical instruments by reflection infrared spectroscopy: A methodological approach. *Anal. Bioanal. Chem.* **2017**, *409*, 3281–3288. [CrossRef]

16. Caruso, F.; Martino, D.F.C.; Saverwyns, S.; Van Bos, M.; Burgio, L.; Di Stefano, C.; Peschke, G.; Caponetti, E. Micro-analytical identification of the components of varnishes from South Italian historical musical instruments by PLM, ESEM–EDX, microFTIR, GC–MS, and Py–GC–MS. *Microchem. J.* **2014**, *116*, 31–40. [CrossRef]

17. Daher, C.; Drieu, L.; Bellot-Gurlet, L.; Percot, A.; Paris, C.; Le Hô, A.-S. Combined approach of FT-Raman, SERS and IR micro-ATR spectroscopies to enlighten ancient technologies of painted and varnished works of art. *J. Raman Spectrosc.* **2014**, *45*, 1207–1214. [CrossRef]

18. Invernizzi, C.; Fiocco, G.; Iwanicka, M.; Kowalska, M.; Targowski, P.; Blümich, B.; Rehorn, C.; Gabrielli, V.; Bersani, D.; Licchelli, M.; et al. Non-invasive mobile technology to study the stratigraphy of ancient Cremonese violins: OCT, NMR-MOUSE, XRF and reflection FT-IR spectroscopy. *Microchem. J.* **2020**, *155*, 104754. [CrossRef]

19. Blümich, B.; Baias, M.; Rehorn, C.; Gabrielli, V.; Jaschtschuk, D.; Harrison, C.; Invernizzi, C.; Malagodi, M. Comparison of historical violins by non-destructive MRI depth profiling. *Microchem. J.* **2020**, *158*, 105219. [CrossRef]

20. Fiocco, G.; Rovetta, T.; Malagodi, M.; Licchelli, M.; Gulmini, M.; Lanzafame, G.; Zanini, F.; Giudice, A.L.; Re, A. Synchrotron radiation micro-computed tomography for the investigation of finishing treatments in historical bowed string instruments: Issues and perspectives. *Eur. Phys. J. Plus* **2018**, *133*, 525. [CrossRef]

21. Fiocco, G.; Rovetta, T.; Gulmini, M.; Piccirillo, A.; Licchelli, M.; Malagodi, M. Spectroscopic Analysis to Characterize Finishing Treatments of Ancient Bowed String Instruments. *Appl. Spectrosc.* **2017**, *71*, 2477–2487. [CrossRef]

22. Bucur, V. *Handbook of Materials for String Musical Instruments*, 1st ed.; Springer International Publishing: Basel, Switzerland, 2016; Chapter 9: The Varnish; pp. 373–453.

23. Chiesa, C.; Hargrave, R.G.; Pollens, S. *Giuseppe Guarneri del Gesù*; Peter Biddulph: London, UK, 1998.

24. Jalovec, K. *Beautiful Italian Violins*; P. Hamlyn: London, UK, 1963.

25. Dondi, P.; Lombardi, L.; Rocca, I.; Malagodi, M.; Licchelli, M. Multimodal workflow for the creation of interactive presentations of 360 spin images of historical violins. *Multimedia Tools Appl.* **2018**, *77*, 28309–28332. [CrossRef]

26. Dondi, P.; Lombardi, L.; Invernizzi, C.; Rovetta, T.; Malagodi, M.; Licchelli, M. Automatic Analysis of UV-Induced Fluorescence Imagery of Historical Violins. *J. Comput. Cult. Heritage* **2017**, *10*, 1–13. [CrossRef]

27. Invernizzi, C.; Fiocco, G.; Iwanicka, M.; Targowski, P.; Piccirillo, A.; Vagnini, M.; Licchelli, M.; Malagodi, M.; Bersani, D. Surface and Interface Treatments on Wooden Artefacts: Potentialities and Limits of a Non-Invasive Multi-Technique Study. *Coatings* **2020**, *11*, 29. [CrossRef]

28. Poggialini, F.; Fiocco, G.; Campanella, B.; Legnaioli, S.; Palleschi, V.; Iwanicka, M.; Targowski, P.; Sylwestrzak, M.; Invernizzi, C.; Rovetta, T.; et al. Stratigraphic analysis of historical wooden samples from ancient bowed string instruments by laser induced breakdown spectroscopy. *J. Cult. Heritage* **2020**, *44*, 275–284. [CrossRef]

29. Dondi, P.; Lombardi, L.; Malagodi, M.; Licchelli, M. 3D modelling and measurements of historical violins. *Acta IMEKO* **2017**, *6*, 29–34. [CrossRef]

30. Gonzalez, S.; Salvi, D.; Baeza, D.; Antonacci, F.; Sarti, A. A data-driven approach to violin making. *Sci. Rep.* **2021**, *11*, 1–9. [CrossRef] [PubMed]

31. Gonzalez, S.; Salvi, D.; Antonacci, F.; Sarti, A. Eigenfrequency optimisation of free violin plates. *J. Acoust. Soc. Am.* **2021**, *149*, 1400–1410. [CrossRef]

32. Fiocco, G.; Rovetta, T.; Invernizzi, C.; Albano, M.; Malagodi, M.; Licchelli, R.A.; Giudice, A.; Gabriele, N.; Lanzafame, G.; Zanini, F.; et al. A Micro-tomographic insight into the coating systems of historical bowed string instruments. *Coatings* **2019**, *9*, 81. [CrossRef]

33. Invernizzi, C.; Rovetta, T.; Licchelli, M.; Malagodi, M. Mid and Near-Infrared Reflection Spectral Database of Natural Organic Materials in the Cultural Heritage Field. *Int. J. Anal. Chem.* **2018**, *2018*, 1–16. [CrossRef]

34. Pellegrini, D.; Duce, C.; Bonaduce, I.; Biagi, S.; Ghezzi, L.; Colombini, M.P.; Tinè, M.R.; Bramanti, E. Fourier transform infrared spectroscopic study of rabbit glue/inorganic pigments mixtures in fresh and aged reference paint reconstruc-tions. *Microchem. J.* **2016**, *124*, 31–35. [CrossRef]

35. Bertrand, L.; Robinet, L.; Cohen, S.X.; Sandt, C.; Le Hô, A.-S.; Soulier, B.; Lattuati-Derieux, A.; Echard, J.-P. Identification of the finishing technique of an early eighteenth century musical instrument using FTIR spectromicroscopy. *Anal. Bioanal. Chem.* **2011**, *399*, 3025–3032. [CrossRef] [PubMed]

36. Stuart, B. *Infrared Spectroscopy: Fundamentals and Applications*; John Wiley & Sons: West Sussex, UK, 2004; ISBN 0470854278.

37. Invernizzi, C.; Daveri, A.; Rovetta, T.; Vagnini, M.; Licchelli, M.; Cacciatori, F.; Malagodi, M. A multi-analytical non-invasive approach to violin materials: The case of Antonio Stradivari "Hellier" (1679). *Microchem. J.* **2016**, *124*, 743–750. [CrossRef]

38. Rosi, F.; Cartechini, L.; Monico, L.; Gabrieli, F.; Vagnini, M.; Buti, D.; Doherty, B.; Anselmi, A.; Brunetti, B.G.; Miliani, C. Tracking metal oxalates and carboxylates on painting surfaces by non-invasive reflection mid-FTIR spectroscopy. In *Metal Soaps in Art*; Casadio, F., Keune, K., Noble, P., van Loon, A., Hendriks, E., Eds.; Springer: Cham, Switzerland, 2019; pp. 173–193.

39. Helwig, K.; Forest, É.; Turcotte, A.; Baker, W.; Binnie, N.E.; Moffatt, E.; Poulin, J. The formation of calcium fatty acid salts in oil paint: Two case studies. In *Metal Soaps in Art*; Casadio, F., Keune, K., Noble, P., van Loon, A., Hendriks, E., Eds.; Springer: Cham, Switzerland, 2019; pp. 173–193.

40.   Rosi, F.; Grazia, C.; Fontana, R.; Gabrieli, F.; Buemi, L.P.; Pampaloni, E.; Romani, A.; Stringari, C.; Miliani, C. Disclosing Jackson Pollock's palette in Alchemy (1947) by non-invasive spectroscopies. *Heritage Sci.* **2016**, *4*, 18. [CrossRef]

41.   Saiz-Jimenez, C. *Molecular Biology and Cultural Heritage*; Routledge: London, UK, 2003.

42.   Canevari, C.; Delorenzi, M.; Invernizzi, C.; Licchelli, M.; Malagodi, M.; Rovetta, T.; Weththimuni, M. Chemical characterization of wood samples colored with iron inks: Insights into the ancient techniques of wood coloring. *Wood Sci. Technol.* **2016**, *50*, 1057–1070. [CrossRef]

Article

Consolidation and Adhesion of Pictorial Layers on a Stone Substrate. The Study Case of the Virgin with the Child from Palazzo Madama, in Turin

Anita Negri [1], Marco Nervo [2,*], Stefania Di Marcello [3] and Daniele Castelli [4]

[1] Science for Conservation, Restoration, Exploitation of Cultural Heritage, University of Turin, 10125 Turin, Italy; anitanegri.1990@gmail.com

[2] Head of Scientific Laboratories, Foundation Centre for Conservation and Restoration "La Venaria Reale", 10078 Venaria Reale, Italy

[3] Art Conservator, Superintendence of Archaeology, Fine Arts and Landscape for the City of L'Aquila, 67100 L'Aquila, Italy; stefania.dimarcello@beniculturali.it

[4] Department of Earth Sciences, University of Turin, 10125 Turin, Italy; daniele.castelli@unito.it

* Correspondence: marco.nervo@centrorestaurovenaria.it

Abstract: The study and the restoration of a polychrome limestone statue representing the Virgin with the Child, from Palazzo Madama in Turin (NW Italy) offered interesting conservation issue related to the polychromy on stone. To preserve the pictorial layers, it was necessary to re-establish the cohesion among the different polychrome layers (original and repainted) and the adhesion between polychrome film and the stone substrate. Particular attention was paid to the choice of intervention materials, selected through a preliminary survey of the scientific literature, and then verified by laboratory tests (tape test, colorimetric test, and permeability test). The most suitable product should to be able to penetrate porosity, to consolidate the layers, to make the pictorial film adhere with the stone surface, and to avoid changes in the colour and in the permeability. The material chosen also had to ensure compatibility with the cleaning method that could only take place after the consolidation of the pictorial layers due to the problematic state of preservation. A range of products, characterised by their small particle size and low viscosity, was tested, and a micro-acrylic resin was selected and successfully applied on the polychromy of the sculpture.

Keywords: polychromy; polychrome stone statue; adhesion and cohesion products; conservation of pictorial films on stone

Citation: Negri, A.; Nervo, M.; Di Marcello, S.; Castelli, D. Consolidation and Adhesion of Pictorial Layers on a Stone Substrate. The Study Case of the Virgin with the Child from Palazzo Madama, in Turin. *Coatings* **2021**, *11*, 624. https://doi.org/10.3390/coatings11060624

Academic Editor: Maude Jimenez

Received: 29 April 2021
Accepted: 20 May 2021
Published: 23 May 2021

Publisher's Note: MDPI stays neutral with regard to jurisdictional claims in published maps and institutional affiliations.

1. Introduction

A polychrome stone consists of a complex and heterogeneous system: the stone substrate and one or more pictorial layers covering the stone. The first is inorganic, porous, transpiring, and hydrophilic; the second ones are less transpiring, partially organic (dyes and several types of binder media), and partially inorganic (mineral pigments). The approach to the consolidation and the adhesion of the pictorial layers covering stone sculptures, therefore, is not uniquely defined.

The object of this study is a polychrome limestone statue representing the Virgin with the Child, from Palazzo Madama in Turin (Figure 1) [1–3]. The work of art has been the subject of a thesis work [4] University of Turin in collaboration with the Foundation Centro Conservazione e Restauro "La Venaria Reale", during the academic year 2017–2018.

Based on comparisons [5–7], the bodies of Mary and Jesus can be attributed to the French style of the 14th century. The head of the Virgin is stylistically similar to some French pieces of the 15th century [8]. The head of the Child is a plaster cast hypothetically made between the 19th and the 20th century, most likely to replace the original missing piece, and it has not been painted. The bodies of the Virgin and the Child are carved in a fine-grained, carbonate sedimentary rock of uncertain provenance, that can be defined as biopelsparite as shown by analysing some samples of the stone under an optical microscope in transmitted polarised light. Mary's head is composed of a carbonate lithotype, a different, more porous and finer-grained limestone. A sample of the the lithotype was observed under an optical microscope and a scanning electron microscope and analysed with an EDX probe. The statue is covered by numerous fragments of pictorial layers, composed of mineral pigments, dyes, and metallic foils characterised by SEM-EDX and FT-IR analyses. The identified inorganic pigments are white lead, gypsum, lead-tin yellow, yellow, brown and red ochre, minium, cinnabar, copper resinate, a different type of a copper-based pigment (verdigris or malachite or chrysocolla), azurite. The analyses also revealed the presence of indigo dye, a red dye and gold and silver foils. For some of the pigments an oil was identified as the binder medium, while in other cases it was no possible to identify the binder by FT-IR analysis. In these cases, FT-IR analyses only revealed the presence of oxalates, together with the pigments. This data led to the hypothesis that the painting technique consisted in a lean tempera (using gums, egg white, calcium dairy or animal glue as binder).

(a)   (b)

Figure 1. The Virgin with the Child from Palazzo Madama in Turin, before (**a**) and after (**b**) the restoration.

Some constituent elements of the pictorial layers are altered or degraded. The presence of oxalates is perhaps due to the hydrolysis of the proteins which compose the binder of some pictorial layers and the consequent interaction of the resulting free fatty acids with metal ions from the pigments, the substrate or the environment [9,10]. The siccative oil is subjected to oxidation, cross-linking and depolymerisation, resulting in the formation of by-products [9,11,12] and loss of cohesion and adhesion. Among the by-products there are groups of free fatty acids, which bind to metal cations coming from the atmosphere or pigments or even from the stone substrate, producing metal soaps [12–14], detected by SEM-EDS analysis. They are subjected to phase changes, resulting in migrations and re-crystallisations within the polychrome layers or in one of the interfaces, forming translucent masses [12,15–17] and causing the formation of bumps on the surface. The analyses revealed the presence of sulphates, probably resulting from the transformation of carbonates (from the stone material, azurite and malachite) by the atmospheric particulate matter and humidity [18]. While the thermo-hygrometric parameters vary, these sulphates are subjected to crystallisation-solubilisation cycles, causing mechanical tensions, which lead to the embrittlement of the pictorial films. Cinnabar is altered in some points, probably due to the photo-oxidation or the interaction with the white lead or with the carbonate of the stone substrate [19,20]. Malachite also is altered, perhaps due to the interaction with soluble salts [20,21]. Finally, the presence of AgCl in silver foils is probably due to chlorides from the atmosphere [22]. The severe state of decohesion and loss of adhesion, required a consolidation and adhesion intervention, since the control of the thermohygrometric parameters was not enough to conserve them.

2. Materials and Methods

To select the most suitable restoration material for this case study, organic (synthetic and natural) and inorganic products were selected on the basis of the evaluation of their characteristics through a bibliographic research and then their testing on mock-ups as representative as possible of the stone-pictorial layers system found for the Virgin with the Child. The tested products are listed in Table 1.

Table 1. Tested products.

Synthetic Organic Resins	Natural Organic Resins	Inorganic Products
25% and 50% Acril ME® (by C.T.S.) [23] in demineralised water and 10% ethanol	1% and 3% Jun Funori (by Lascaux Colours & Restauro) in demineralised water [24–29]	2% and 5% Ammonium oxalate (by Sinopia) in demineralised water [30,31]
100% Acrylmat (by AN.T.A.RES) [23]	rabbit skin glue (by C.T.S.) in a ratio of 1:14 in water [24,25]	100% Nanolime CaLoSiL® E25 (by IBZ-Salzchemie GmbH & Co. KG) [32]
8% and 16% Dispersion® K52 (by Kremer Pigmente) in demineralised water [33–35]	sturgeon glue (by AN.T.A.RES) in a ratio of 1:20 in water [25]	-
33% and 50% Micral-Primal WS-24® (by Kremer Pigmente) in demineralised water	-	-
5% and 10% Microacril® CV40 (by Chem Spec) in demineralised water [36]	-	-
3% and 10% Aquazol® 200 (by Polymer Chemistry Innovations) in demineralised water and 10% ethanol [25,37–40]	-	-
100% Fluoline® HY (by C.T.S.) [41]	-	-
10% and 20% Regalrez® 1126 (by C.T.S.) in ligroin [24,42]	-	-
2% and 4% Tylose® MH 300p (by C.T.S.) in demineralised water [37]	-	-

Blocks (5 cm × 5 cm × 1 cm) were prepared with lime putty and limestone dust in a ratio of 1:3. Vinci limestone powder was selected as an aggregate: it is entirely composed of calcium carbonate and was sieved to select aggregates with ≤200–300 μm^3 dimensions (Figure 2). A plaster rasp was used to imitate the tooth chisel processing, obtaining a material which was similar to the constituting stone of the bodies of the Virgin and the Child, from a chemical and morphological point of view, as verified by an optical microscope. Since the polychrome stratigraphy of the work of art is very complex and extremely heter-ogeneous, a simplification was implemented to realise the mock-ups, even though in our opinion their representativeness was not compromised. Therefore, rather than reproducing all the layers of each area of the sculpture, only two pictorial layers were prepared. One was composed of red ochre, white lead and siccative oil (linseed oil), the other consisted of yellow ochre, white lead, and animal glue. The two pictorial layers were spread by brush on sheets of Melinex®; once dried, they were peeled off and subjected to artificial aging. During the present case study, the artificial aging in the stove lasted twenty-two days: for eleven days the temperature was kept at 60 °C and then at 110 °C for another eleven days. Following this treatment, it was possible to observe a notable stiffening of the pictorial layers and a slight variation in the colours.

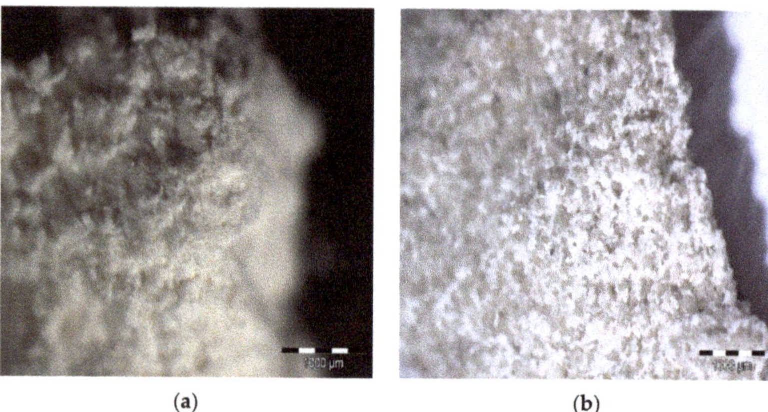

(a)   (b)

Figure 2. (a) One of the mortar samples seen under the optical microscope, with 25× magnification. (b) Sample of lithotype from the base of the bodies of the Virgin and the Child, seen under the optical microscope, with 25× magnification.

The adhesive power, which is considered to be the most important requirement in this case study, was evaluated with tape tests.

Each product tested was applied on a mock-up composed by the block representing the lithotype, covered by an oil painting layer and, on top, one tempera layer, both laid in fragments. The tested products were applied by syringe, allowing them to penetrate between the layers (Figure 3). The effectiveness of the organic resins was evaluated three to four days after the first application; in the case of a visible inefficacy, the product was applied a second time. Inorganic consolidants were applied four to six times. Regarding the case of nanolime, the tape test was performed thirty days after its application. Regarding each product to be tested, three pieces of Scotch® 3M were applied in three different points on the surface of each sample and then de-attached. After this, the average weight difference between the piece of tape after and before the application was calculated, using an Acculab ALC-210.4 analytical balance (Acculab, New York, NY, USA), with readability 0.1 mg. The lower the resulting value, the greater the adhesive strength of the product. The mock-ups were photographed before and after the test (Table A1, Appendix A), to see and record which layers adhered and, consequently, to evaluate the penetration capacity of the products among the different interfaces.

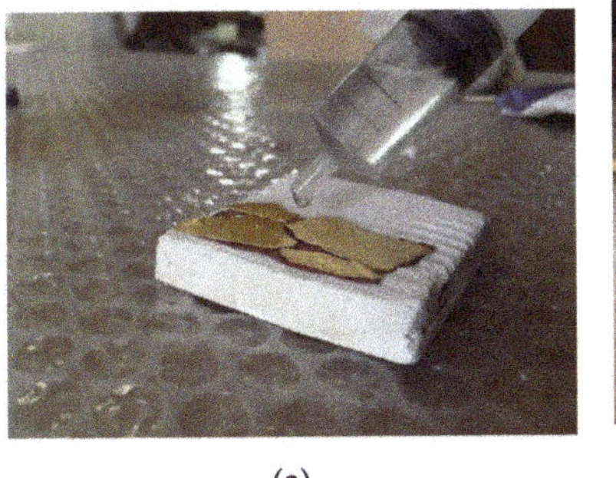 

(a)   (b)

Figure 3. (**a**) Application of a resin upon a sample. (**b**) After the application of the product, on each specimen a weight was placed, with the interposition of a Melinex® sheet, to increase the adhesion between the layers.

The products resulting with a satisfactory adhesion strength were then subjected to the colorimetric test. The Italian recommendation NorMaL 43/93 defines the colorimetric coordinates collection on opaque surfaces of stone materials which are subjected to alterations and conservative treatments. According to NorMaL 43/93 it is possible to perform a quantitative measurement of the colour variation by detecting the colorimetric coordinates before and after the application of the product. During the present case study, a Konica-Minolta CM-700d colorimeter (Tokyo Metropolis, Japan) was used, which was equipped with a pulsed xenon lamp and an integrating sphere for diffused illumination of the sample (geometry d/8, illuminant D65, measurement conditions SCI + SCE, measurement step 10 nm, spectral range 400–700 nm). The acquisitions were performed on an oil painting film (prepared as above described). Each product was applied with a brush in a single coat. After drying, three colorimetric measurements of three distinct points were conducted for each sample.

Finally, the most suitable product must not form a water vapour impermeable film. To perform a water vapour permeability test, reference was made to the European standard EN ISO 12572 (June 2001 edition), which describes a method for determining the hygroscopic permeance of building products and materials in an environment characterised by constant temperature and relative humidity. The test was performed on samples consisting of a waterproof box containing water, with an opening, entirely covered by a mortar block (prepared as already described). Each product to be tested was applied in two layers by brush on the surface of three blocks, trying to obtain a uniform and similar application for all mock-ups. Due to the different partial pressure of the water vapour between the box and the external environment, the vapour passed through the blocks of stone material. Via periodically weighing the system, it was possible to detect the flow of water vapour that passed through the porosity of the material, determining its hygroscopic permeance. During this case study we did not want to acquire absolute measurements, but relative data, thus, to compare the difference in permeability of a porous material following a treatment with resins.

3. Results

The adhesive strength, which is considered to be the most important requirement in this case study, has been evaluated with tape test. Then, the products with a satisfying adhesive strength have been subjected to colorimetric tests, to evaluate the possible change in colour, and to permeability tests, to evaluate the variation of this parameter in the stone substrate after consolidation.

3.1. Adhesion Test

Seen in Table 2 and in Table A1 in Appendix A, with 10% and 20% Regalrez®, rabbit glue, CaLoSil®, and 5% ammonium oxalate, no adhesion was obtained, so performing the tape tests was deemed unproductive. The 3% and 10% Aquazol® 200, 5% Microacril® CV40, sturgeon glue 1:20, and 2% ammonium oxalate did not get a satisfactory adhesion between the layers. A better, but not sufficient, result was achieved with Micral-Primal® WS-24 at 33% and 50%, Acryl ME® at 25% and pure Fluoline® HY (applied twice), and with Microacril® CV40 at 10% (applied once). The 50% Acryl ME®, pure AcrilMat®, 8% and 16% Dispersion® K52, 3% Funori, 2% and 4% Tylose® MH300 showed a good adhesive strength. Nevertheless, it should be noted that the Acryl ME® and pure AcrilMat® were effective only after two applications. The first application the products likely penetrated within the porosity too in depth. Despite the effectiveness of the second application, it was decided to discard these two resins for the application on the sculpture, because their excessive penetrability makes the cohesion/adhesion operation difficult to control. The 50% Acril ME®, 5% Microacril® CV40, and 2% ammonium oxalate achieved good adhesion between the two pictorial layers, but not between these latter and the stone substrate.

Table 2. Tape test. Difference was calculated as weight after the test and before the test.

Sample Name	Number of Applications of the Consolidant	Difference of Weight (mg)	Average Weight of the Scotch Tape after Tape Test (mg) *
25% Acryl ® ME	2	243 47 57	116 ± 98
50% Acryl® ME	2	33 56 49	46 ± 11
Acrylmat®	2	0 0 0	0 ± 0
3% Aquazol® 200	2	796 - -	no adhesion
10% Aquazol® 200	2	344 552 -	no adhesion
16% Dispersion® K52	1	0 0 0	0 ± 0
8% Dispersion® K52	1	0 0 0	0 ± 0
Fluoline® HY	2	132 177 12	107 ± 82
33% Micral-Primal® WS-24	1	3 4 0	2 ± 1
50% Micral-Primal® WS-24	1	4 0 0	1 ± 1
1% Jun Funori	2	4 1 11	5 ± 5
3% Jun Funori	2	0 1 0	0 ± 0

Table 2. *Cont.*

Sample Name	Number of Applications of the Consolidant	Difference of Weight (mg)	Average Weight of the Scotch Tape after Tape Test (mg) *
10% Regalrez® 1126	1	- - -	no adhesion
20% Regalrez 1126	1	- - -	no adhesion
5% Microacril® CV40	1	762 - -	no adhesion
10% Microacril® CV40	1	0 0 337	112 ± 168
2% Tylose® MH 300p	1	1 0 0	0 ± 0
4% Tylose® MH 300p	1	0 0 0	0 ± 0
2% Ammonium Oxalate	1	0 846 -	no adhesion
5% Ammonium Oxalate	1	- - -	no adhesion
Rabbit glue Lapin	4	- - -	no adhesion
Sturgeon glue	4	748 - -	no adhesion
CaLoSil®	6	- - -	no adhesion

* The mean absolute error was calculated as semi-difference between the higher and the lower value, considering the limited number of measurements. This method allows not to underestimate the error.

The adhesives that showed satisfactory efficacy on the samples—2% Tylose® MH300p, 3% Jun Funori and 8% and 16% Dispersion® K52—therefore, were selected and subjected to the following tests. The 4% Tylose MH300p was excluded, despite its good adhesive strength, as its density makes it difficult to distribute and it causes colour saturation evident to the naked eye.

3.2. Colorimetric Test

Seen in Tables 3–6, among the tested products, 16% Dispersion® K52 caused a greater variation of the b^* coordinate: this implies a slight yellowing of the surface, although it is not perceptible to the naked eye. The ΔE_{00} of 8% Dispersion® K52 and of 3% Jun Funori slightly exceeded the value 1, so the chromatic variation was negligible.

Table 3. Colorimetric test: Average ΔE_{00}.

Resin	Average ΔE_{00}
16% Dispersion® K52	0.99 ± 0.32
8% Dispersion® K52	1.90 ± 0.42
3% Jun Funori	1.13 ± 0.24
2% Tylose® MH300	0.86 ± 0.67

Table 4. Colorimetric test: average ΔL^*(D65).

Resin	Average ΔL^*(D65)
16% Dispersion® K52	0.45 ± 0.45
8% Dispersion® K52	0.34 ± 0.55
3% Jun Funori	0.36 ± 0.20
2% Tylose® MH300	0.84 ± 0.75

Table 5. Colorimetric test: average Δa^*(D65).

Resin	Average Δa^*(D65)
16% Dispersion® K52	0.45 ± 0.38
8% Dispersion® K52	0.01 ± 0.20
3% Jun Funori	0.18 ± 0.26
2% Tylose® MH300	0.57 ± 0.17

Table 6. Colorimetric test: average Δb^*(D65).

Resin	Average Δb^*(D65)
16% Dispersion® K52	0.62 ± 1.13
8% Dispersion® K52	0.46 ± 0.18
3% Jun Funori	0.02 ± 0.38
2% Tylose® MH300	0.18 ± 0.22

The products that caused a minor colour variation are 2% Tylose MH300p and 8% Dispersion® K52.

3.3. Permeability Test

The purpose of this test is to compare the permeability of a porous material following the treatment with an intervention product. Regarding the case of a polychrome stone sculpture, the artistic technique itself leads to a reduction in permeability, since the pictorial layers tend to be less porous than the stone substrate. It is appropriate to not worsen this condition, applying where necessary a product that does not further occlude the porosity, leaving the actual level of permeability as much unaltered as possible.

As can be seen in Table 7, even though the number of samples is limited and the standard deviation is high, the test has an indicative value. It is possible to note that there were no significant differences in behaviour of the products tested, and each of them reduced permeability by approximately 20%.

Table 7. Permeability test.

Resin	Total Weight Average of Evaporated Water Vapour (mg)
No resin	985 ± 42
16% Dispersion® K52	781 ± 63
8% Dispersion® K52	746 ± 58
3% Jun Funori	764 ± 38

4. Discussion

The ideal product must achieve the cohesion of the pictorial films and the adhesion between them and the stone substrate. It must not change the gloss nor colour of the pictorial surface, it must guarantee the durability over time of both physical features and colour, avoiding by-products that would be a direct or indirect cause of alteration in the future. It must allow the retractability of the intervention. Reversibility is not considered a priority since, in the future the possibility that it will be considered necessary to remove one or more pictorial layers is remote and this operation would be possible mechanically, as demonstrated by tests on mock-ups. Therefore, we also tested inorganic consolidants, which have excellent stability over time. The ideal product must be compatible with the chosen surface cleaning methods. During our case study, different water cleaning processes were selected, after tests were performed to remove consistent deposits from the different materials of which the sculpture is made. The consolidation study started when it was not possible to select the suitable cleaning method yet; for this reason, water-sensitive products were initially included.

The critical conservative state of the pictorial layers required us to achieve the consolidation and the adhesion of polychromy simultaneously or, in some cases, even before the surface cleaning. A pre-consolidation would have fixed the substantial coherent deposits, while a preliminary cleaning would have compromised the conservation of the pictorial films. A Long Q-Switch (LQS) laser (El.En. S.p.a., Florence, Italy) was tested on some detached fragments, as this method implies low mechanical stress, but it caused the alteration of cinnabar and white lead and the partial removal of this latter. Q-Switch (QS), Short Free Run (SFR) and Erbium-YAG lasers were not tested. Some case studies [43–45] show that the first (QS laser) causes the blackening of cinnabar and white lead. The second (SFR laser) acts mainly by spallation, leading to an increase in temperature and pressure, with the risk of causing excessive mechanical stress. The Erbium-YAG laser does not induce the alteration of the white lead and cinnabar, likely due to the presence of an organic binder, which protects the pigments [46]. However, in the present study, as the pictorial layers on the sculpture are very poor in binder, its use probably would have failed. Although the possibility of cleaning by temporarily fixing the polychrome layers with cyclododecane in ligroin [24,36,47,48] also was considered, this method was found unsuitable for achieving a satisfactory level of cleaning. After tests, to re-establish the adhesion of the ochre films, which contain clay minerals, an attempt was made through moisturising (to eliminate electrostatic forces), a slight heating and pressure exerted with a thermocautery, and interposing Melinex® [12]. This process was assumed to avoid the contraindications of a pre-consolidation and the introduction of substances which are extraneous to the artwork, but it was unsuccessful in our case study. To be shown in Section 5. Conclusion, the optimal way to clean and consolidate the pictorial films consisted of introducing the selected product, waiting for it to start to polymerise and cleaning, exploiting the "softening" action released by the water of the applied emulsion. A water cleaning gel (gellan gum [49,50]) was selected for the stone surfaces, characterised by the presence of little and scattered pictorial fragments: in this case, the punctual consolidation was performed before the surface cleaning.

To select the most suitable one, some inorganic and organic (synthetic and natural) products were selected based on their viscosity and particle size, two characteristics that allow the resin to penetrate through and among the layers. To know these data, we referred to the technical data sheets of the manufacturers which, however, are often incomplete. The products whose at least one of these two parameters is indicated in the data sheets were selected and those having a viscosity ≤ 1000 mPa·s and a particle diameter ≤ 0.1 μm were tested. Polyvinyl acetate-based polymers were not taken into account because they may hydrolyse releasing acetic acid, which is harmful to the lithotype. In pictorial films, hydrolysis of the ester is favoured by temperature and humidity and could be eventually catalysed by metal cations or oxides coming from pigments [37]. Synthetic products traditionally used on polychrome stone sculptures, such as Paraloid®, were not tested, as they do not meet the viscosity and particle size requirements. An exception was made for the rabbit skin glue and the sturgeon glue, which are natural products commonly used in restoration, particularly in the field of paintings on wood, and for Funori, which was studied and applied for the adhesion of polychromy on stone also [24,51]. These resins have the advantage of being easily removable over time without using organic solvents, and of having a low toxicity.

Following the tests, it was decided to use the acrylic microemulsion Dispersion® K52 for the consolidation of the pictorial layers that cover the stone surfaces of the Virgin and the Child, thanks to its excellent adhesive strength, the absence of significant chromatic variations and its comparability with other products in terms of permeability.

Jun Funori provided excellent results and has a better stability to UV radiation than the acrylic resins [24,25,27,42]. Nevertheless, the cleaning tests performed on the Virgin with the Child revealed the suitability of the aqueous methods, which are incompatible with Funori. Regarding the average photosensitivity of the acrylic resins [39,52], we must consider that the selected microemulsion has a good penetration capacity, thus reducing this shortcoming; also, the risk of the presence of residues on the statue surfaces was reduced by cleaning with cotton swabs during the application phase. We finally must consider that the sculpture is conserved in a museum environment. It, therefore, is possible and necessary to monitor not only the thermohygrometric parameters, but also the exposure to light, avoiding irradiation with UV rays. These conditions also are essential to prevent the alteration of some constituent materials of the work of art itself, such as cinnabar, which are sensitive to light.

It seems clear that the adhesion of pictorial layers on a stone substrate causes a reduction in permeability. The water can leave a deposit of salts under the treated surface, which can rehydrate and recrystallize at any thermo-hygrometric variation. These phenomena lead to mechanical stresses within the porosity, causing breakages at the interfaces and the decohesion of materials [53]. The moisture that accumulates below the impermeable surface also can react with the constituent materials of the polychrome sculpture, such as the silver foil and the lithotype, thus participating in the formation of carboxylates [12]. Over the centuries, in addition to the siccative oil and the probable lean tempera, silver and golden foils were applied to decorate the work of art. Consequently, it is certainly necessary to deepen this research, to formulate more suitable products. Future research must focus not only on the compatibility, stability over time, retractability, good adhesive strength, and absence of chromatic alteration, but also on the permeability variation, caused by the introduction of an adhesive/consolidating agent in a complex and heterogeneous system.

5. Conclusions

The application of Dispersion® K52 on the pictorial films of the sculpture confirmed the good results that emerged from the tests. The resin was applied by syringe at 8 and 16% in demineralised water over the pictorial layers which were in a state of decohesion and loss of adhesion, after the application of water and ethanol at 50%. This allowed us to lower the surface tension and to partially wash away the deposits which were accumulated under the pictorial layers. Regarding more localised applications, the same product was used at 10% in isopropyl alcohol [35], avoiding the use of water and ethanol at 50% and the consequent dispersion. The introduction of the microemulsion hydrated the coherent deposits, allowing us to easily remove them with swabs or scalpels, reaching a satisfying cleaning level, without affecting the polychromy, thanks to the initial crosslinking of the resin. The microemulsion also softened the pictorial layers, allowing us to lay down the pictorial scales on the stone substrate, exerting a slight and gradual pressure with a spatula and swabs, and interposing Melinex®. After that, the pictorial layers were more legible (Figure 4).

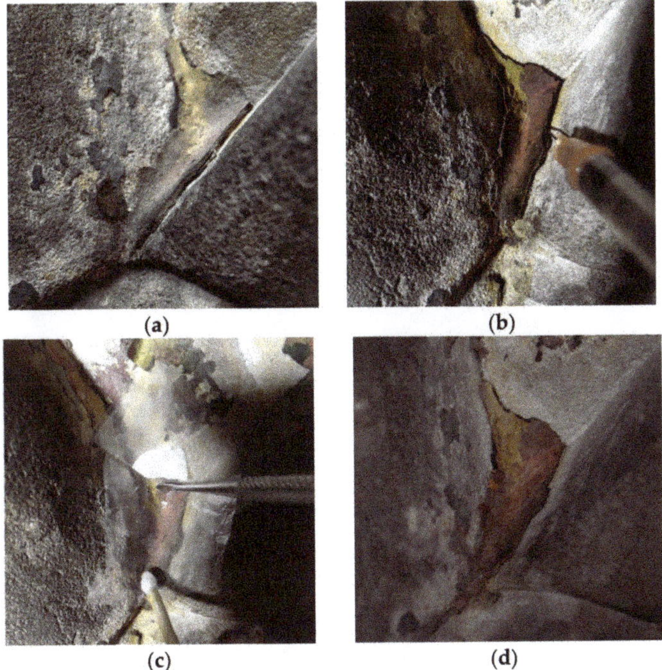

Figure 4. (**a**) The pictorial layers before the consolidation by syringe (**b**), cleaning with buffers and exerting a gradual pressure to adhere the films to the substrate (**c**). (**d**) After the consolidation.

Author Contributions: Conceptualisation, D.C., S.D.M., A.N. and M.N.; methodology, D.C., S.D.M., A.N., M.N.; validation, S.D.M., A.N. and M.N.; formal analysis, M.N.; resources, Centro Conservazione e Restauro "La Venaria Reale" in collaboration with Università degli Studi di Torino; data curation, A.N., M.N.; writing—original draft preparation, A.N; writing—review and editing, A.N., M.N.; visualisation, D.C., S.D.M., A.N. and M.N.; supervision, M.N. All authors have read and agreed to the published version of the manuscript.

Funding: This research received no external funding.

Institutional Review Board Statement: Not applicable.

Informed Consent Statement: Not applicable.

Data Availability Statement: Data is contained within the article.

Acknowledgments: We would like to thank Soprintendenza delle Belle Arti e Paesaggio per il Comune e la provincia di Torino, especially the art conservator Valeria Moratti and the conservator of Museo Civico d'Arte Antica di Torino Simone Baiocco for their kind permission.

Conflicts of Interest: The authors declare no conflict of interest.

Appendix A

Table A1. The specimens subjected to tape test.

Before Tape Test	After Tape Test

25% and 50% Acril® ME	25% and 50% Acril® ME

AcrilMat®	AcrilMat®

3% and 10% Aquazol® 200	3% and 10% Aquazol® 200

Table A1. *Cont.*

Before Tape Test	After Tape Test
 16% Dispersion® K52	 16% Dispersion® K52
 8% Dispersion® K52	 8% Dispersion® K52
 Fluoline® HY	 Fluoline® HY

Table A1. *Cont.*

Before Tape Test	After Tape Test
33% and 50% Micral-Primal® WS-24	33% and 50% Micral-Primal® WS-24
1% and 3% Jun Funori	1% and 3% Jun Funori
10% and 20% Regalrez® 1126	10% and 20% Regalrez® 1126

Table A1. *Cont.*

Before Tape Test	After Tape Test
5% and 10% Microacril® CV40	5% and 10% Microacril® CV40
2% and 4% Tylose® MH 300p	2% and 4% Tylose® MH 300p
2% and 5% ammonium oxalate	2% and 5% ammonium oxalate

Table A1. *Cont.*

Before Tape Test	After Tape Test
CaLoSil®	CaLoSil®
Rabbit glue Lapin and sturgeon glue	Rabbit glue Lapin and sturgeon glue

References

1. Mallé, L. Recenti acquisti di sculture antiche al Museo Civico di Torino. *Cronache Econ.* **1969**, *322*, 3.
2. Mallé, L. *I Musei Civici di Torino. Acquisti e Doni 1966–1970. Catalogo Museo Civico di Torino*; Galleria Civica d'Arte Moderna: Torino, Italy, 1970; p. 5.
3. Mallé, L. *Palazzo Madama in Torino, II. Le Collezioni d'Arte*; Tipografia Torinese Editrice: Torino, Italy, 1970.
4. Negri, A. Vergine Col Bambino Proveniente da Palazzo Madama: Studio E Intervento per la Conservazione Della Policromia E Il Recupero della Leggibilità Dell'Opera. Master's Thesis, Università Degli Studi di Torino, Turin, Italy, 2019.
5. Baron, F. *Sculpture Francaise. I. Moyen Age*; Rèunion des Musées Nationaux: Paris, France, 1996.
6. Baron, F.; Avril, F.; Chapu, P.; Gaborit-Chopin, D.; Perrot, F. *Les Fastes du Gothique: Le Siècle de Charles V*; Rèunion des Musées Nationaux: Paris, France, 1981.
7. Taburet-Delahaye, E.; Avril, F. Paris 1400: Les arts sous Charles VI, Paris. In Proceedings of the Musee du Louvre, Editions de la Réunion des Musées Nationaux, Paris, France, 22 March–12 July 2004; pp. 216–335.
8. Vv.Aa. Paris, 1400: Les arts sous Charles VI. Musée du Louvre. In Proceedings of the Fayard, Réunion des Musées Nationaux, Paris, France, 22 March–12 July 2004; pp. 216–335.
9. Mills, J.S.; White, R. *The Organic Chemistry of Museum Objects*; Butterworths: London, UK, 1994.
10. Matteini, M.; Moles, A. *La Chimica nel Restauro. I Materiali dell'Arte Pittorica*; Nardini Editore: Firenze, Italy, 2010.
11. Pizzini, S. Indagine Sul Comportamento Ossidativo di Film Pittorici ad Olio di Produzione Industriale Mediante Ozonizzazione e Caratterizzazione GC/MS. Master's Thesis, Università Ca' Foscari, Venice, Italy, 2012.
12. Aguado-Guardiola, E. Estudio Del Rol de Los Agregados Minerales en la Formación, Envejecimiento y Conservación de Películas Pictóricas Al Óleo. Ph.D. Thesis, Universitat Politècnica de València, Valencia, Spain, 2017.

13. Mecklenburg, M.F.; Tumosa, C.S.; Erhardt, D. Structural response of painted wood surfaces to changes in ambient relative humidity. In Proceedings of the Painted Wood: History and Conservation, a symposium organized by the Wooden Artifacts Group of the American Institute for Conservation of Historic and Artistic Works and the Foundation of the AIC, Williamsburg, VA, USA, 11–14 November 1994; pp. 464–483.

14. Aguado-Guardiola, E.; Fuster-Lòpez, L. The role of stone substrate in oil paint film stability: An insight into some issues influencing durability and conservation. In Proceedings of the of Polychrome Sculpture: Decorative Practice and Artistic Tradition, ICOM-CC Interim Meeting, Working Group Sculpture, Polychromy, and Architectural Decoration, Tomar, Portugal, 28–29 May 2013; pp. 16–26.

15. Boon, J.J.; Keune, K.; Zucker, J. Imaging analytical studies of lead soaps aggregating in preprimed canvas used by the Hudson River School painter F.E. Church. *Microsc. Microanal.* **2005**, *11*, 444–445. [CrossRef]

16. Keune, K.; Boon, J.J. Analytical imaging studies of cross-sections of paintings affected by lead soap aggregate formation. *Stud. Conserv.* **2007**, *52*, 161–176. [CrossRef]

17. Boon, J.J.; Keune, K.; Van der Weerd, J.; Noble, P.; Wadum, J. Mechanical and chemical changes in Old Master paintings: Dissolution, metal soap formation and remineralization processes in lead pigmented ground/intermediate paint layers of 17th Century paintings. In Proceedings of the 13th Triennial Meeting of the ICOM Committee for Conservation, Rio de Janerio, Brazil, 22–27 September 2002.

18. Torraca, G. *Lectures on Materials Science for Architectural Conservation*; The Getty Conservation Institute: Los Angeles, CA, USA, 2009.

19. Rinaldi, S.; Quartullo, G.; Milaneschi, A.; Pietropaoli, R.; Occorsio, S.; Costantini Scala, F.; Minunno, G.; Virno, C. *La Fabbrica dei Colori: Pigmenti e Coloranti Nella Pittura e Nella Tintoria*; Il bagatto: Roma, Italy, 1995.

20. Bevilacqua, N.; Borgioli, L.; Adrover Garcia, I.; Matteini, M. *I Pigmenti Nell'Arte: Dalla Preistoria Alla Rivoluzione Industriale*; Il Prato: Vicenza, Italy, 2010.

21. Roy, A. *Artists' Pigments: A Handbook of Their History and Characteristics*; National Gallery of Art: Washington, DC, USA, 1993.

22. Lyon-Marion, B.A.; Mittelman, A.M.; Rayner, J.; Lantagne, D.S.; Pennell, K.D. Impact of chlorination on silver elution from ceramic water filters. *Water Res.* **2018**, *142*, 471–479. [CrossRef] [PubMed]

23. Borgioli, L.; Camaiti, M.; Rosi, L. Comportamento all'irraggiamento UV di nuovi formulati polimerici per il restauro. In Proceedings of the VI Congresso Annuale IGIIC, Spoleto, Italy, 2–4 October 2008.

24. Prestipino, G.; Santamaria, U.; Morresi, F.; Amenta, A.; Greco, C. Sperimentazione di adesivi e consolidanti per il restauro di manufatti lignei policromi egizi. In Proceedings of the Lo Stato dell'Arte, XIII Congresso Nazionale IGIIC, Centro Conservazione e Restauro La Venaria Reale, Torino, Italy, 22–24 October 2015.

25. Kunzelman, D. L'attenzione alle superfici pittoriche. Materiali e metodi per il consolidamento e metodi scientifici per valutarne l'efficacia. In Proceedings of the the the IV Congresso Internazionale Colore e Conservazione, Materiali e Metodi nel Restauro Delle Opere Policrome Mobili, Milano, Italy, 21–22 November 2008.

26. Geiger, T.; Michel, F. Studies on the polysaccharide JunFunori used to consolidate matt paint. *Stud. Conserv.* **2005**, *50*, 193–204. [CrossRef]

27. Finozzi, A. *Progetto Restauro. Speciale n. 62. Utilizzo Della Colla Funori Nel Restauro*; Il Prato: Saonara, Italy, 2006.

28. Llamas Pacheco, R.; San Pedro, D.R. Colorimetric evaluation of three adhesives used in the consolidation of contemporary matte paint after artificial ageing. *Conserv. Patrimònio* **2014**, *20*, 11–21. [CrossRef]

29. Harrold, J.; Wyszomirska-Noga, Z. Funori: The use of a traditional Japanese adhesive in the preservation and conservation treatment of Western objects. In *Adapt & Evolve 2015: East Asian Materials and Techniques in Western Conservation, Proceedings of the International Conference of the Icon Book & Paper Group, London, UK, 8–10 April 2015*; The Institute of Conservation: London, UK, 2017; pp. 68–79.

30. Matteini, M. Inorganic treatments for the consolidation and protection of stone artefacts and mural paintings. *CSCH* **2008**, *8*, 13–27. [CrossRef]

31. Borgioli, L. *Protezione e Consolidamento con "Bario Idrato" ed "Ammonio Ossalato"*; Technical Report; CTS: Hopkinton, MA, USA, 2012; pp. 1–7.

32. Di Gaetano, S. Restauro di Due Sculture Lapidee con Tracce di Policromia: Studio di Ricomposizione Non Invasivo. Master's Thesis, Università Degli Studi di Torino, Turin, Italy, 2015.

33. Genova, I. Il Restauro del Bozzetto Scultoreo del Carro di San Rocco di Pietro Consagra. Uno Studio Sul Comportamento Chimico-Fisico di Leganti Pittorici Impiegati Nell'Arte Contemporanea. Master's Thesis, Università Degli Studi di Palermo, Palermo, Italy, 2015.

34. Puglisi, C.; Reginella, M.L.; Sottile, S.; Bruno, M.; Vitella, M. Restoration of a polymateric sculpture of the immaculate conception. *Eur. J. Sci. Theol.* **2017**, *13*, 69–77.

35. Rubino, C. Definizione Delle Metodologie di Intervento Su Un Dipinto Murale di Età Romana Proveniente Dall' Area Vesuviana: Sistemi Tradizionali ed Innovativi a Confronto Per Il Recupero Della Superficie Pittorica. Master's Thesis, Università Degli Studi di Torino, Turin, Italy, 2018.

36. Pelottieri, M. Il Restauro di Una Scultura in Marmo Dipinta del Museo Civico D'Arte Antica di Torino: Studio e Progettazione di Un Metodo di Pulitura e Consolidamento. Master's Thesis, Università Degli Studi di Torino, Turin, Italy, 2018.

37. Borgioli, L.; Cremonesi, P. *Le Resine Sintetiche Usate nel Trattamento di Opere Policrome*; Il Prato: Saonara, Italy, 2005.

38. Wolbers, R.C.; McGynn, M.; Duerbeck, D. Poly(2-Ethyl-2-Oxazoline): A new conservation consolidant. In *Painted Wood: History and Conservation, Proceedings of the Symposium in Williamsburg, November 11–14, 1994*; Getty Conservation Institute: Los Angeles, CA, USA, 1994; pp. 514–527.

39. Semenzato, C. Studio e Analisi del Degrado di Leganti Polimerici Utilizzati per il Ritocco Pittorico. Master's Thesis, Università Ca' Foscari Venezia, Venice, Italy, 2015.

40. Arslanoglu, J. Evaluation of the use of Aquazol as an adhesive in painting conservation. *WAAC Newsl.* **2003**, *25*, 2.

41. Borgioli, L. *Polimeri di Sintesi per la Conservazione della Pietra*; Il Prato: Saonara, Italy, 2006.

42. Castelli, G.; Gigli, M.C.; Lalli, C.; Lanterna, G.; Weiss, C.; Speranza, L. Un composto organico sintetico per il consolidamento del legno: Sperimentazione, misure e prime applicazioni. *OPD Restauro* **2002**, *14*, 144–152.

43. Pouli, P.; Emmony, D.C.; Madden, C.E.; Sutherland, I. Analysis of the laser-induced reduction mechanisms of medieval pigments. *Appl. Surf. Sci.* **2001**, *173*, 252–261. [CrossRef]

44. Cooper, M.I.; Fowles, P.S.; Tang, C.C. Analysis of the laser-induced discoloration of lead white pigment. *Appl. Surf. Sci.* **2002**, *201*, 75–84. [CrossRef]

45. Pustka, M.L.; Violini, P.; Cencia, F.; Ferlito, A.; Giacomazzi, P.; Heiniger, C.; Leopardi, F.; Munzi, C.; Pinto, G.; Sechi, S.; et al. La rimozione di ridipinture a biacca alterate sugli affreschi della cappella di San Lorenzo alla Scala Santa. In Proceedings of the Aplar 5. Applicazioni Laser Nel Restauro, Musei Vaticani, Vatican, 18–20 September 2014.

46. Korenberg, C.; Pereira-Pardo, L. The use of erbium lasers for the conservation of cultural heritage. A review. *J. Cult. Herit.* **2018**, *31*, 236–247. [CrossRef]

47. Anselmi, C.; Presciutti, F.; Doherty, B.; Daveri, A.; Miliani, C.; Brunetti, B.G.; Sgamellotti, A. Ottimizzazione dei metodi di applicazione del ciclododecano come protettivo temporaneo per interventi in emergenza. In Proceedings of the Lo Stato dell'Arte, VI Congresso Nazionale IGIIC, Spoleto, Italy, 2–4 October 2008.

48. Rowe, S.; Rozeik, C. The uses of cyclododecane in conservation. *Stud. Conserv.* **2008**, *53*, 17–31. [CrossRef]

49. Cremonesi, P. *L'ambiente Acquoso per il Trattamento di Opere Policrome*; Il Prato: Saonara, Italy, 2012.

50. Placido, M. Il Restauro e la Protezione della Carta Mediante Trattamento con Gel di Gellano. Master's Thesis, Università Sapienza, Rome, Italy, 2019.

51. Gérard-Bendelé, A.; Le Pogam, P.Y. Un retable du XIV siècle (musée de Bar-le-Duc): La complémentarité de la sculpture et de la polychromie. *Techné* **2014**, *39*, 87–89.

52. Clifton, J.R. *Stone Consolidating Materials: A Status Report*; U.S. Department of Commerce, National Bureau of Standards: Washington, DC, USA, 1980.

53. Casoli, A.; Cauzzi, D.; Palla, G. Lo studio dei leganti in opere policrome: Gli oli siccativi. *OPD Restauro* **1999**, *11*, 115–117.

Article

Characterization and Identification of Varnishes on Copper Alloys by Means of UV Imaging and FTIR

Miriam Truffa Giachet *, Julie Schröter * and Laura Brambilla *

Haute Ecole Arc Conservation Restauration, HES-SO University of Applied Sciences and Arts Western Switzerland, 2000 Neuchâtel, Switzerland

* Correspondence: miriam.truffagiachet@gmail.com (M.T.G.); julie.schroter@he-arc.ch (J.S.); laura.brambilla@he-arc.ch (L.B.); Tel.: +41-32-930-19-19

Abstract: The application of varnishes on the surface of metal objects has been a very common practice since antiquity, both for protective and aesthetic purposes. One specific case concerns the use of tinted varnishes on copper alloys in order to mimic gilding. This practice, especially flourishing in the 19th century for scientific instruments, decorative objects, and liturgical items, results in large museum collections of varnished copper alloys that need to be preserved. One of the main challenges for conservators and restorers deals with the identification of the varnishes through non-invasive and affordable analytical techniques. We hereby present the experimental methodology developed in the framework of the LacCA and VERILOR projects at the Haute École ARC of Neuchâtel for the identification of gold varnishes on brass. After extensive documentary research and analytical campaigns on varnished museum objects, various historic shellac-based varnishes were created and applied by different methods on a range of brass substrates with different finishes. The samples were then characterized by UV imaging and infrared spectroscopy before and after artificial ageing. The comparative study of these two techniques was performed for different thicknesses of the same varnish and for different shellac grades in order to implement an identification methodology based on simple non-invasive examination and analytical tools, which are accessible to conservators.

Keywords: varnishes; shellac; copper alloys; UV-induced fluorescence; FTIR spectroscopy; eddy current; UV imaging; artificial ageing; non-invasive characterization

Citation: Truffa Giachet, M.; Schröter, J.; Brambilla, L. Characterization and Identification of Varnishes on Copper Alloys by Means of UV Imaging and FTIR. *Coatings* **2021**, *11*, 298. https://doi.org/10.3390/coatings11030298

Academic Editor: Nervo Marco

Received: 19 February 2021
Accepted: 3 March 2021
Published: 5 March 2021

Publisher's Note: MDPI stays neutral with regard to jurisdictional claims in published maps and institutional affiliations.

1. Introduction

The application of coatings on metal surfaces, especially on copper alloys, is an ancient procedure already in practice since antiquity [1]. Decorative objects, liturgical items, and scientific instruments in particular were often varnished mainly as a protection against corrosion and for esthetic purposes. As copper alloys can naturally exhibit a more or less yellow hue, it is easy to give a golden shine to the surface by applying a tinted transparent coating on the metal substrate. The application of gold varnishes on decorative bronze objects in order to mimic gilding is attested since the 17th century in France [2], although Theophilus mentions the application of a yellow varnish on tin decorative leaves already in the 12th century [3]. The so-called "gilt bronzes", a misleading appellative referring most of times to brass alloys, are hence very common in museum collections and they need to be preserved. In reason of the optical illusion created by gold varnishes on the surface, one of the main challenges for conservators is the ability to systematically discriminate them from genuine gold layers without the use of complex techniques of characterization. Moreover, the surface is sometimes too strongly worn off to come to a hasty conclusion.

It is in this framework that LacCa and VERILOR projects came about at the Haute École ARC of Neuchâtel. The LacCA project was dedicated to the elaboration of a protocol of identification and characterization of lacquered copper alloys using simple affordable methods, accessible to conservators, as well as sophisticated characterization methods. The current project VERILOR focuses, on the other hand, on gold varnishes found on decorative bronze objects dating from the 19th century. This project includes the possible presence of restoration materials on the surfaces in order to integrate the identification and characterization protocol outlined in LacCa. Furthermore, historical aspects are explored more in-depth in order to provide a better understanding of the cultural significance of these surface finishes and to be able to preserve them more efficiently.

An extensive 19th century literature review provided information about the materials and methods used to manufacture the objects. Statistics on gold lacquers recipes show that alcohol-based varnishes containing shellac and dyes were the most commonly used to imitate gilding on copper alloys. Shellac is a naturally brownish resin secreted by the female lac bug on trees in the forests of India and comes in various grades depending on the level of purity, bleaching, and dewaxing [4]. The shellac-based varnishes were tinted using a high variety of natural dyes, such as turmeric, sandarac, elemi, saffron, dragon's blood, as well as the first synthetic aniline-based colorants in the second half of the 19th century. These varnishes were then filtered to obtain perfectly transparent coatings.

The investigation of these recipes was firstly approached by the study of pure shellac films of different grades before exploring more complex mixtures including other binding media and colorants. The aim of the article is to study the effectiveness of standard examinations techniques used by conservators (i.e., UV imaging), as well as more complex characterization methods (i.e., FTIR—Fourier-transform infrared spectroscopy) in order to identify pure shellac-based varnishes on copper alloys. The correlation between the two techniques is explored for different grades of shellac varnishes applied in various thicknesses on mock-up samples, including the artificial ageing of the coupons as well. The general overview and methodological aspects of part of this study were presented during the ICOM-CC metals working group conference in 2019 and published in [5].

The examination of objects under UV light is one of the classic, simple, and practical investigation techniques used by conservators and restorers to verify the presence of fluorescent materials non detectable under white light [6–8]. Although databases including the UV fluorescence of materials exist [4,9–11], few studies have been done specifically on gold varnishes applied on copper alloys from the 19th century [12–14]. As for infrared spectroscopy, this technique has proven to be excellent in the identification of complex organic compounds, as well as in the discrimination of different organic materials used as coatings in art [15,16]. Studies of varnishes on metal substrate by FTIR include [13,17,18].

2. Materials and Methods

2.1. Sample Preparation

Various experimental series of samples were created in the framework of the LacCA and VERILOR projects exploring the identification of transparent varnishes on copper alloys. The variables taken into account were the type and the finish of the metal substrate, the nature of the varnish applied, the varnish application method, and the thickness of the varnished layer. The characteristics of the 24 samples selected for comparison are listed in Table 1.

The selected brass coupons are made from CuZn36 and CuZn37 alloys and they were treated to obtain two surface finishes, namely mirror-polish and brush finish ("satin").

Various varnish recipes were selected based on extensive bibliographic research from a repertory of over 100 recipes [5]; all selected varnishes are shellac-based because of their high occurrence in the 19th century literature; ethanol was used as solvent for the same reason. Different grades of shellac were chosen for comparison (Table 2), namely bleached dewaxed shellac in flakes, orange shellac in flakes (both with wax and dewaxed), and dewaxed seedlac in grains by Kremer Pigmente, as well as three industrial bleached dewaxed shellac products, i.e., "Astra" from Laverdure (in flakes), "Astra" from Boesner (liquid), and "Platina" from Laverdure (in flakes). In addition, a sample of pure shellac wax from Kremer Pigmente was created by melting the wax directly on the coupon on a hot plate.

Table 1. Characteristics of the samples analyzed in this study. Varnish recipes are presented in Table 2. Sample code interpretation: 1st letter = type of substrate finish (S = satin, M = mirror-polished); 2nd (and 3rd) letter = coating technique (md = manual dip-coating; id = industrial dip-coating; b = brush; m = melting); letter between underscore and dash = varnish recipe; after the dash (when present) = to discriminate between samples' duplicates.

Sample Code	Substrate Finish	Varnish Recipe	Coating Technique	Number of Layers	Artificial Ageing (Hours)
Smd_A	Satin		Manual dip-coating	1	-
Mmd_A	Mirror-polish	A	Manual dip-coating	1	-
Mb_A	Mirror-polish		Brush	1	-
Sb_A	Satin		Brush	1	-
Mid_A-2	Mirror-polish		Industrial dip-coating	1	-
Mid_A-11	Mirror-polish		Industrial dip-coating	1	-
Smd_B	Satin	B	Manual dip coating	1	-
RM	Mirror-polish	-	Not varnished	-	-
RS	Satin	-	Not varnished	-	-
Sb_1L-a	Satin	1L	Brush	1	1512
Sb_1L-b					-
Sb_1L-c	Satin	1L	Brush	2	1512
Sb_1L-d					-
Sb_2L-a	Satin	2L	Brush	1	1512
Sb_2L-b					-
Sb_2L-c	Satin	2L	Brush	2	1512
Sb_2L-d					-
Sb_8V-a	Satin	8V	Brush	1	1415
Sb_8V-b					-
Smd_1L-ast	Satin	1L	Manual dip-coating	1	1415
Smd_AS	Satin	AS *	Manual dip-coating	1	1415
Smd_1L-pla	Satin	1L	Manual dip-coating	1	1415
RS-1	Satin	-	Not varnished	-	1415
Sm_SW	Satin	SW **	Melting	1	-

* "Astra" shellac by Boesner, liquid and ready-to-use (not listed in Table 2, see main text). ** Industrial shellac wax, not diluted in ethanol (not listed in Table 2, see main text).

Table 2. Varnish recipes listed in Table 1.

Ingredients	A	B	1L	2L	8V
Bleached dewaxed shellac	125 g	-	225 g	-	-
Orange non-dewaxed shellac	-	225 g	-	-	-
Orange dewaxed shellac	-	-	-	225 g	-
Dewaxed seedlac	-	-	-	-	63 g
Ethanol	2 kg	1 L	1 kg	1 kg	1 L

The varnishes were created by dissolving the shellac in ethanol in a water bath at 50 °C on a heating magnetic stirrer. Different ethanol grades were used depending on the original experimental goals. Coupons varnished with recipes A and B were in fact created to evaluate the spectral response and the thicknesses of the purest shellac varnish possible, and they were therefore fabricated with 99% grade ethanol. The other samples were created to replicate the historical recipes as accurately as possible, and they were therefore produced using less pure ethanol (96%). After preparation, the solutions were vacuum filtered using Büchner flask and funnel and Whatman paper filters (grade 602 h). Recipe B orange shellac was not dewaxed, whereas recipe 2L orange shellac and seedlac were dewaxed using the traditional decanting technique to obtain a completely transparent varnish. The solutions were left several days to decant, and the upper clear liquid was then carefully transferred into clean bottles to be stored in the dark. The Kremer Pigmente, Laverdure, and Boesner bleached shellac, on the contrary, were already industrially dewaxed.

Three coating methods were employed to apply the varnishes on the coupons: manual dip coating, brush application, and industrial dip coating. The first two traditional methods, widely mentioned in the literature [19,20], were chosen to simulate the thinnest varnish layer possible. On the other hand, automated dip coating was performed in order to obtain a homogenous coating layer of known thickness. The details of these application methods can be found in [5]. Two layers of varnish were applied by brush on some sample (Table 1) in order to simulate thicker coatings that could be found on historical objects. Before varnishing, the coupons surface was treated in two ways: the satin coupons destined for recipes A and B and for shellac wax and the mirror-polished coupons were degreased with ethanol with a soft cloth, whereas the other coupons were scraped with powdered pumice stone by means of a toothbrush, rinsed with tap water, then left for few minutes in a bath of sulfuric acid 0.1 M, rinsed again and finally degreased in ethanol and dried. This procedure was performed in order to eliminate possible oxidation products present on the surface, according to procedures found in the literature [21]. As the samples were not heavily tarnished, the coupons were cleaned in one step and not stripped heavily before using several chemical solutions as it is done for freshly cast brass ornaments.

After coating, the samples were left to dry for few days away from contamination and then stocked in transparent boxes to be stored in the dark.

2.2. Artificial Ageing

Some of the samples were placed in a climate chamber in order to evaluate possible changes of the varnishes with the ageing. Natural resins, in fact, have shown changes in some of their physical and chemical characteristics with ageing, as for example in their type of fluorescence under UV light [22,23]. It was therefore chosen to submit some of the samples to accelerated weathering by UV radiation in order to simulate the damaging light exposure conditions to which historic decorative objects and scientific instrument might be exposed indoors.

The samples were placed for 2 months (the total number of hours for each sample is indicated in Table 1) in a Memmert ICH L climate chamber (Memmert GmbH + Co. KG, Büchenbach, Germany) at 50% relative humidity and 25 °C temperature to replicate the average indoor environmental conditions. The two Sylvania Blacklight BL368 UV tubes (Feilo Sylvania, Budapest, Hungary) placed on the upper part of the chamber emit in the UVA spectral range (315–400 nm), with an emission peak at 368 nm. These values are in line with the ISO exposure recommendation for paintings and varnishes [24] and they simulate the portion of the UV daylight not being filtered by common glass windows. The coupons were placed side by side to cover the whole surface of the chamber and they were arranged in order to have the best reproducibility in exposure.

2.3. Characterization Methods

2.3.1. Thickness Measurements

The thickness of the samples was measured with a Phynix Surfix Pro S gauge and FN 1.5 eddy-current probe (PHYNIX GmbH & Co. KG, Neuss, Germany) with a measurement range of 0–1.5 mm and an accuracy, with foil calibration, of ± 1.0 μm + 1% of value. All measurements were taken in the same areas of each sample by means of paper masks created according to the size of the coupon. The accuracy of this technique was assessed through a comparative study with spectroscopic ellipsometry and confocal microscopy on the samples varnished with recipes A and B [5].

2.3.2. UV Imaging

Imaging under UV light was performed on all samples with an unfiltered Canon EOS 750D camera (Canon INC, Tokyo, Japan) with an 18–35 mm lens, a Baader UV/IR cut (Baader Planetarium, Munich, Germany) and an X-Nite CC1 filter (LDP LLC, Carlstadt, NJ, USA). The samples, placed on a black non-UV emitting cardboard background, were illuminated by two Dutscher UV lamps (Dominique DUTSCHER SAS, Bernolsheim, France) with emission peak at 365 nm mounted at 45° on a custom-made stand. An UV Innovations target was used as reference in order to later calibrate the pictures in Adobe Photoshop. The pictures were taken in manual mode, ISO 200 and f/11 aperture; the target was photographed with 1 s exposure time, whereas the samples with 30 s exposure time. This choice was made in order to be able to see the fluorescence of even the thinnest layers of varnish, otherwise not visible with shorter exposure times. White balance correction was performed on the pictures in Adobe Photoshop Camera Raw following the Adobe Photoshop Setup and Capture Workflows recommended by the target manufactures (https://www.uvinnovations.com/getting-started (accessed on March 2018). No exposure correction was applied to the samples. $L^* a^* b^*$ values of three spots (11 px × 11 px) were then recorded for each coupon in Adobe Photoshop using the color sampling tool in order to compare the color of the fluorescence of each varnish.

2.3.3. Fourier-Transform Infrared Spectroscopy

All samples were analyzed with a ThermoFischer Scientific Nicolet iS 5 FTIR spectrometer (ThermoFischer Scientific, Waltham, MA, USA) in reflectance mode. Spectra were collected using 128 scans at 4 or 8 cm^{-1} resolution, measuring between 4000 and 650 cm^{-1}. A custom-made external module was used to be able to compare the coupons spectra with those obtained from museum objects analyzed with the same configuration (Figure 1). Coupons varnished with recipe 8V were also analyzed with a benchtop ThermoFischer Scientific Nicolet iN 10 MX FTIR microscope (ThermoFischer Scientific, Waltham, MA, USA) with the same collection parameters in order to have better spectral response.

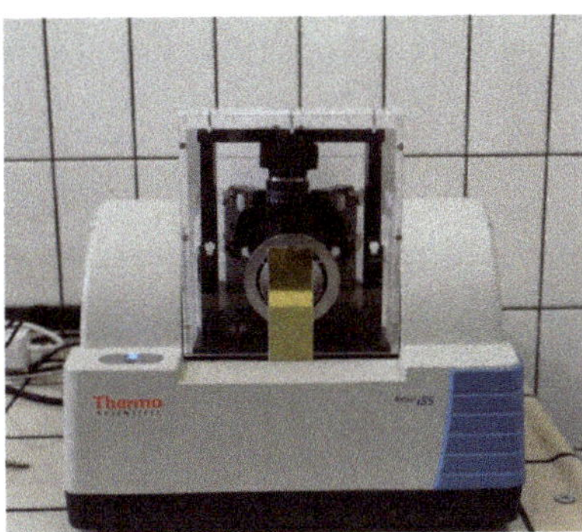

Figure 1. The FTIR instrument with custom-made external module used to analyze the samples.

Baseline correction and atmospheric suppression were applied to the raw FTIR spectra.

3. Results

3.1. Thickness Measurement

Table 3 shows the average thickness of the samples measured with an eddy-current probe.

Table 3. Average thickness of the samples obtained by eddy-current probe. Standard deviation at 1σ.

Code	Average Thickness (μm)
Smd_A	0.7 ± 0.3
Mmd_A	1.0 ± 0.2
Mb_A	0.7 ± 0.2
Sb_A	0.9 ± 0.2
Mid_A-2	2.1 ± 0.3
Mid_A-11	11.2 ± 0.6
Smd_B	1.3 ± 0.2
Sb_1L-a	3.3 ± 0.2
Sb_1L-b	2.8 ± 0.3
Sb_1L-c	5.0 ± 0.2
Sb_1L-d	3.8 ± 0.2
Sb_2L-a	1.3 ± 0.1
Sb_2L-b	1.4 ± 0.2
Sb_2L-c	2.1 ± 0.3
Sb_2L-d	1.8 ± 0.4
Sb_8V-a	1.4 ± 0.5
Sb_8V-b	1.4 ± 0.3
Smd_1L-ast	1.4 ± 0.1
Smd_AS	11.8 ± 0.4
Smd_1L-pla	1.3 ± 0.2
Sm_W	50.3 ± 13.9

Free-hand measurements with eddy-current probe resulted to be precise after adequate calibration with thin foil [5].

3.2. UV Imaging and Colorimetry

Figure 2 shows the fluorescence under UV light and the *L***a***b** colorimetric coordinates of the different types of shellac varnishes, the shellac wax, and the not-varnished sample. Samples Sb_A, Sb_1L-b, Sb_8V-a, and Sb_2L-b were varnished by brush, whereas varnish on samples Smd_1L-ast, Smd_1L-pla, and Smd_AS was applied by manual dip-coating.

Figure 3 shows how samples varnished with recipe A, 1L, and 2L, sorted by mode of varnish application and thickness of the layer applied, appear under UV light in comparison with non-varnished reference coupons; the *L** *a** *b** colorimetric coordinates of the fluorescence of the varnishes on different supports and with different thicknesses obtained in Adobe Photoshop are listed as well.

Figure 4 shows how samples with different varnishes appear under UV light before and after artificial ageing, taking into account the *L** *a** *b** colorimetric coordinates and the thickness of the varnishes. Due to the improvement and the changes of the UV imaging setup during the experiments, it was not possible to compare the same coupon before and after ageing for varnishes 1L and 2L. A non-aged duplicate of each type was therefore used as the "before ageing" sample.

Bleached dewaxed shellac	Bleached dewaxed shellac	Bleached dewaxed shellac	Bleached dewaxed shellac	Bleached dewaxed shellac
Recipe A	Recipe 1L	Recipe 1L	Recipe 1L	Ready-to-use
Sb_A	**Sb_1L-b**	**Smd_1L-ast**	**Smd_1L-pla**	**Smd_AS**
0.9 ± 0.2 µm	2.8 ± 0.3 µm	1.4 ± 0.1	1.3 ± 0.2	11.8 ± 0.4
$L* = 11 \pm 1$	$L* = 11 \pm 1$	$L* = 9 \pm 1$	$L* = 6 \pm 1$	$L* = 46 \pm 1$
$a* = 4 \pm 1$	$a* = 6 \pm 1$	$a* = 7 \pm 1$	$a* = 4 \pm 1$	$a* = -7 \pm 0$
$b* = 4 \pm 1$	$b* = 5 \pm 1$	$b* = 3 \pm 1$	$b* = 4 \pm 1$	$b* = 0 \pm 1$
Dewaxed seedlac	Orange dewaxed shellac	Orange non-dewaxed shellac	Shellac wax	Not-varnished sample
Recipe 8V	Recipe 2L	Recipe B	Ready-to-use	-
Sb_8V-a	**Sb_2L-b**	**Smd_B**	**Sm_W**	**RS-1**
1.4 ± 0.1	1.4 ± 0.2 µm	1.3 ± 0.2	50.3 ± 13.9	-
$L* = 10 \pm 2$	$L* = 17 \pm 1$	$L* = 42 \pm 9$	$L* = 72 \pm 6$	$L* = 2 \pm 0$
$a* = 12 \pm 0$	$a* = 15 \pm 1$	$a* = 25 \pm 3$	$a* = -9 \pm 1$	$a* = 3 \pm 1$
$b* = 10 \pm 2$	$b* = 17 \pm 1$	$b* = 30 \pm 5$	$b* = 12 \pm 2$	$b* = 0 \pm 0$

Figure 2. Results of the UV imaging and the colorimetric analysis performed on samples with different grades of shellac.

3.3. Fourier-Transform Infrared Spectroscopy

Figure 5 shows the FTIR spectra of the representative samples of each varnish recipe: samples Mb_A (recipe A), Sb_1L-d (recipe 1L), Smd_1L-ast ("Astra"), Smd_1L-pla ("Platina"), and Smd_AS ("Astra", ready-to-use) for bleached dewaxed shellac; samples Smd_B (recipe B) and Sb_2L-d (recipe 2L) for orange shellac; sample Sb_8V-b (recipe 8V) for seedlac; Sm_W for shellac wax.

Figure 6 shows the FTIR spectra of samples before and after ageing. Spectra Sb_8V-a and Sb_8V-b were acquired with the benchtop instrument in order to obtain better spectral response (cf. 2.3.3).

Bleached dewaxed shellac		Bleached dewaxed shellac	
Recipe A, manual dip-coating		Recipe A, industrial dip-coating	
Mmd_A	**Smd_A**	**Mid_A-2**	**Mid_A-11**
1.0 ± 0.2 μm	0.7 ± 0.3 μm	2.1 ± 0.3 μm	11.2 ± 0.6 μm
$L^* = 6 \pm 1$	$L^* = 7 \pm 1$	$L^* = 13 \pm 0$	$L^* = 33 \pm 1$
$a^* = 2 \pm 1$	$a^* = 5 \pm 1$	$a^* = 4 \pm 0$	$a^* = 7 \pm 1$
$b^* = -6 \pm 0$	$b^* = -2 \pm 1$	$b^* = 5 \pm 1$	$b^* = 12 \pm 1$
Bleached dewaxed shellac		Bleached dewaxed shellac	
Recipe A, brush application		Recipe 1L, brush application	
Mb_A	**Sb_A**	**Sb_1L-b**	**Sb_1L-d**
0.7 ± 0.2 μm	0.9 ± 0.2 μm	2.8 ± 0.3 μm	3.8 ± 0.2 μm
$L^* = 10 \pm 1$	$L^* = 11 \pm 1$	$L^* = 11 \pm 1$	$L^* = 19 \pm 1$
$a^* = 2 \pm 1$	$a^* = 4 \pm 1$	$a^* = 6 \pm 1$	$a^* = 5 \pm 1$
$b^* = 5 \pm 0$	$b^* = 4 \pm 1$	$b^* = 5 \pm 1$	$b^* = 7 \pm 1$
Orange dewaxed shellac		**Not-varnished references**	
Recipe 2L, brush application			
Sb_2L-b	**Sb_2L-d**	**RM**	**RS**
1.4 ± 0.2 μm	1.8 ± 0.4 μm	-	-
$L^* = 17 \pm 1$	$L^* = 25 \pm 1$	$L^* = 2 \pm 1$	$L^* = 2 \pm 0$
$a^* = 15 \pm 1$	$a^* = 18 \pm 1$	$a^* = 4 \pm 0$	$a^* = 4 \pm 0$
$b^* = 17 \pm 1$	$b^* = 23 \pm 1$	$b^* = 1 \pm 1$	$b^* = 0 \pm 1$

Figure 3. Results of the UV imaging and the colorimetric analysis performed on varnishes with different thicknesses and on different supports.

Bleached dewaxed shellac		Bleached dewaxed shellac		Orange dewaxed shellac	
Varnish 1L		Varnish 1L		Varnish 2L	
 Before	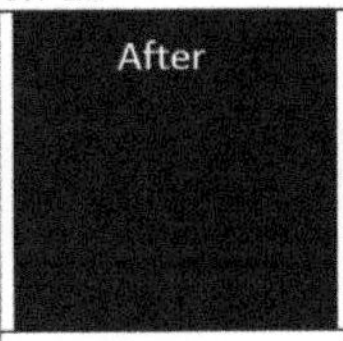 After	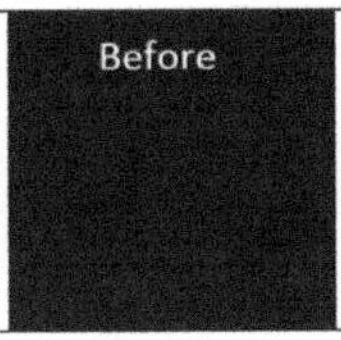 Before	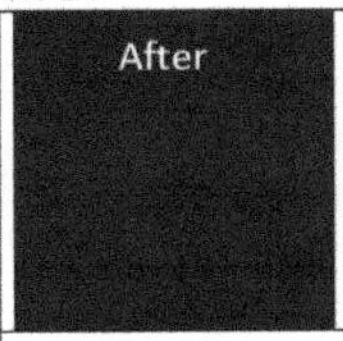 After	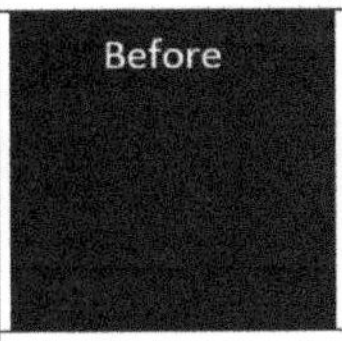 Before	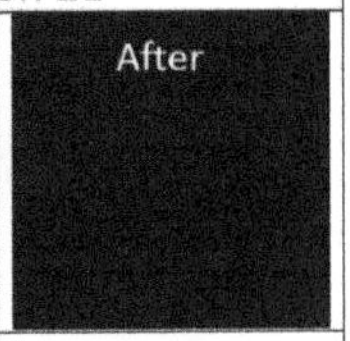 After
Sb_1L-b	**Sb_1L-a**	**Sb_1L-d**	**Sb_1L-c**	**Sb_2L-b**	**Sb_2L-a**
2.8 ± 0.3 µm	3.3 ± 0.2 µm	3.8 ± 0.2 µm	5.0 ± 0.2 µm	1.4 ± 0.2 µm	1.3 ± 0.1 µm
$L^* = 11 \pm 1$	$L^* = 12 \pm 0$	$L^* = 19 \pm 1$	$L^* = 24 \pm 1$	$L^* = 17 \pm 1$	$L^* = 14 \pm 1$
$a^* = 6 \pm 1$	$a^* = 3 \pm 1$	$a^* = 5 \pm 1$	$a^* = 0 \pm 0$	$a^* = 15 \pm 1$	$a^* = 13 \pm 1$
$b^* = 5 \pm 1$	$b^* = -2 \pm 0$	$b^* = 7 \pm 1$	$b^* = -4 \pm 0$	$b^* = 17 \pm 1$	$b^* = 13 \pm 1$

Orange dewaxed shellac		Dewaxed seedlac		Bleached dewaxed shellac	
Varnish 2L		Varnish 8V		Varnish 1L	
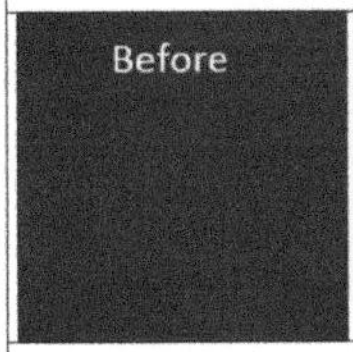 Before	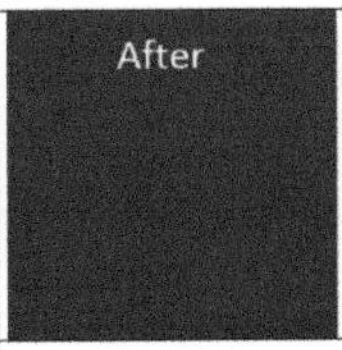 After	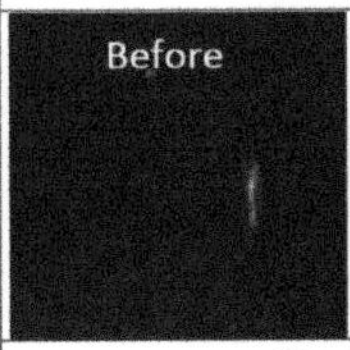 Before	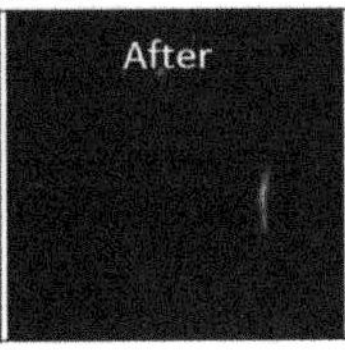 After	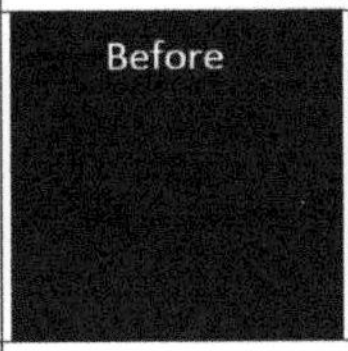 Before	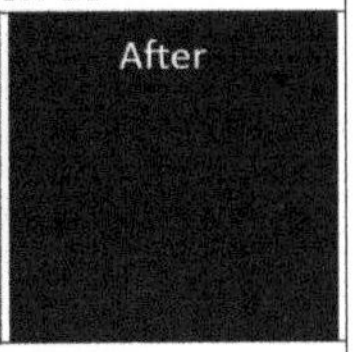 After
Sb_2L-d	**Sb_2L-c**	**Sb_8V-a**		**Smd_1L-pla**	
1.3 ± 0.1 µm	2.1 ± 0.3 µm	1.4 ± 0.5 µm		1.3 ± 0.2 µm	
$L^* = 25 \pm 1$	$L^* = 25 \pm 0$	$L^* = 10 \pm 2$	$L^* = 9 \pm 1$	$L^* = 6 \pm 1$	$L^* = 7 \pm 1$
$a^* = 18 \pm 1$	$a^* = 18 \pm 0$	$a^* = 12 \pm 0$	$a^* = 9 \pm 1$	$a^* = 4 \pm 1$	$a^* = 4 \pm 1$
$b^* = 23 \pm 1$	$b^* = 23 \pm 0$	$b^* = 10 \pm 2$	$b^* = 8 \pm 2$	$b^* = 4 \pm 1$	$b^* = 4 \pm 1$

Bleached dewaxed shellac		Bleached dewaxed shellac		Not-varnished reference	
Varnish ready-to-use		Varnish 1L			
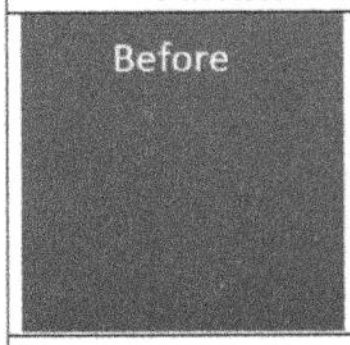 Before	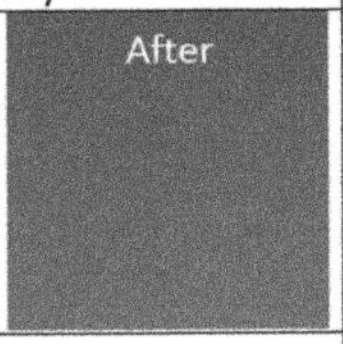 After	 Before	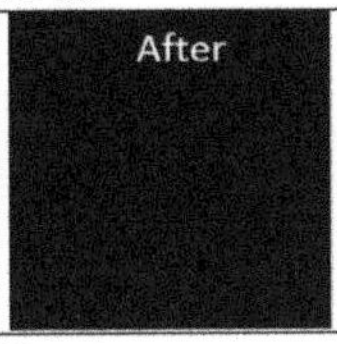 After	 Before	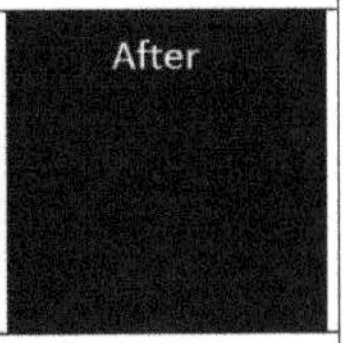 After
Smd_AS		**Smd_1L-ast**		**RS-1**	
11.8 ± 0.4 µm		1.4 ± 0.1 µm		-	
$L^* = 46 \pm 1$	$L^* = 50 \pm 1$	$L^* = 9 \pm 1$	$L^* = 9 \pm 1$	$L^* = 2 \pm 0$	$L^* = 2 \pm 1$
$a^* = -7 \pm 0$	$a^* = -7 \pm 1$	$a^* = 7 \pm 1$	$a^* = 6 \pm 1$	$a^* = 3 \pm 1$	$a^* = 4 \pm 1$
$b^* = 0 \pm 1$	$b^* = 0 \pm 1$	$b^* = 3 \pm 1$	$b^* = 2 \pm 1$	$b^* = 0 \pm 0$	$b^* = 0 \pm 0$

Figure 4. Results of the UV imaging and the colorimetric analysis performed on samples before and after artificial ageing.

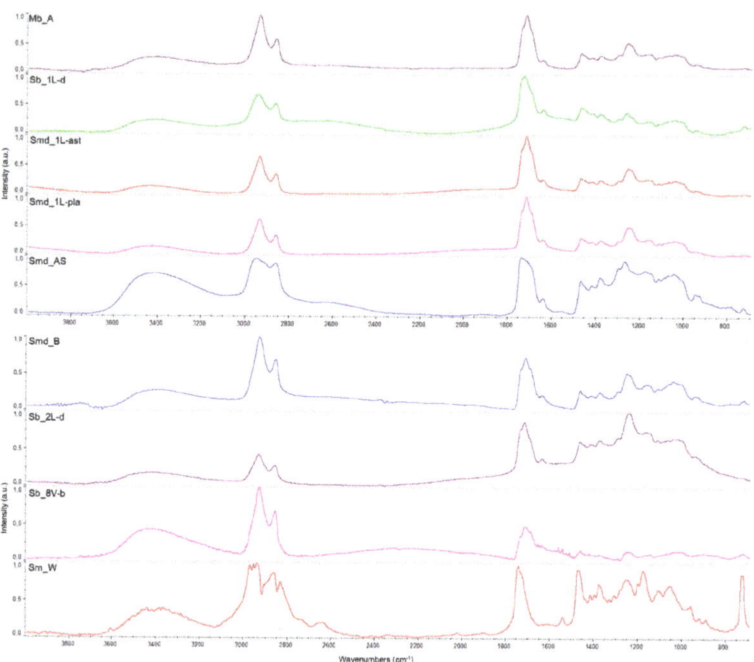

Figure 5. FTIR spectra of the representative samples for each varnish recipe and shellac wax. Spectrum Sb_8V-b was acquired with the benchtop instrument (cf. 2.3.3).

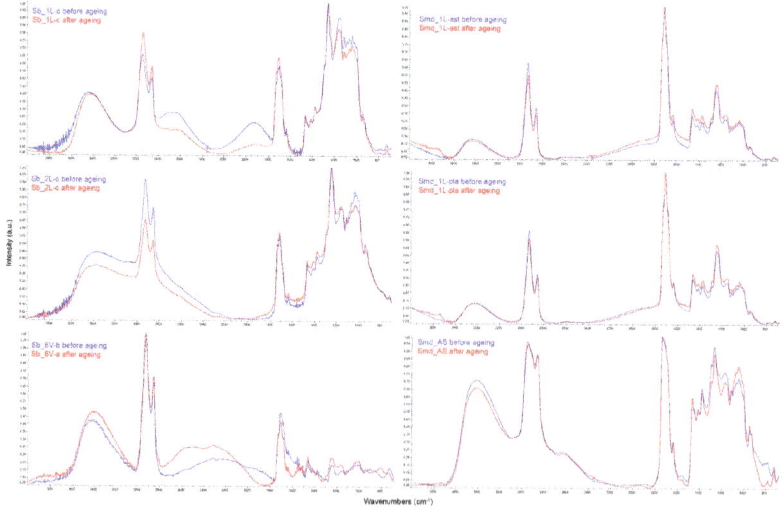

Figure 6. FTIR spectra of the samples before (blue spectra) and after (red spectra) ageing. On the left, from top to bottom: Sb_1L-c, Sb_2L-c, Sb_8V-b (and Sb_8V-a); on the right, from top to bottom: Smd_1L-ast, Smd_1L-pla, Smd_AS.

4. Discussion

4.1. Comparison between Different Grades of Shellac

As shown in Figures 2 and 7, different grades of shellac can be differentiated by their fluorescence under UV light.

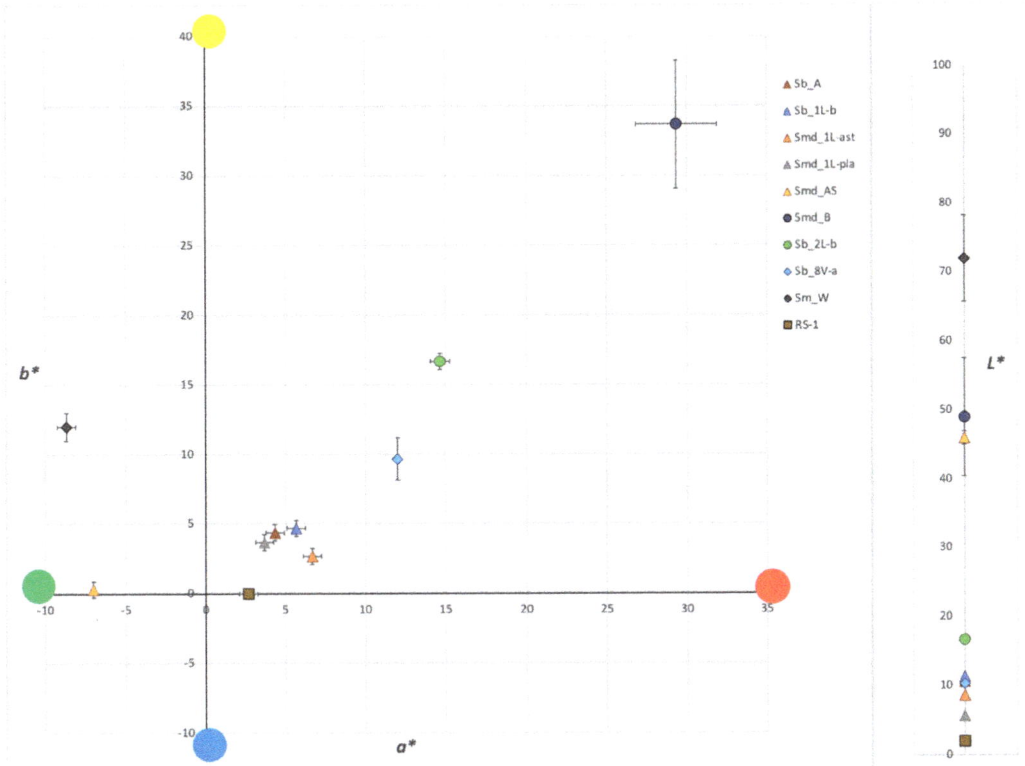

Figure 7. *L* a* b** colorimetric coordinates of the varnishes' fluorescence under UV light (cf. Figure 2). ▲: bleached dewaxed shellac; ●: orange shellac with or without wax; ◆: seedlac and shellac wax; ■: non-varnished references.

It is possible to observe that orange shellac and seedlac exhibit a quite visible orange fluorescence, which is stronger when wax is present in the varnish (sample Smd_B): shellac wax is in fact very fluorescent under UV light, as corroborated by the aspect of the pure shellac wax sample Sm_W. On the other hand, bleached dewaxed shellac shows a weak fluorescence tending towards neutral tinges, with the exception of the ready-to-use "Astra" shellac by Boesner (sample Smd_AS), exhibiting a much greener hue. This fluorescence color is usually linked to other natural vegetal resins, such as mastic, dammar, or rosin [4,10,25,26]. The latter is in fact present in traces in this industrial product, as indicated in its ingredients list. This might explain why the FTIR spectrum of the Smd_AS shellac sample is the only one presenting slight variations in comparison to all other pure shellac samples (Figure 8).

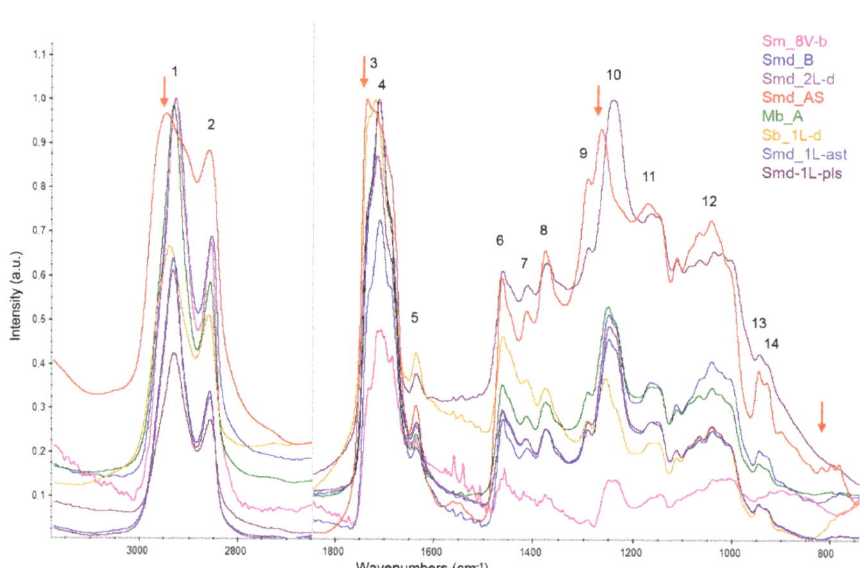

Figure 8. Difference between sample Smd_AS and the other shellac samples analyzed (red arrows). The number of top of the peaks (1−14) refers to peak assignments reported in Table 4.

Table 4. Characteristic peaks of shellac. Peak assignment after [27,28].

Number in Figure 8	Molecular Motion	Wavelength (cm^{-1})	Number in Figure 8	Molecular Motion	Wavelength (cm^{-1})
1	C–H stretch	2934–2920	8	C–H bend	1377
2	C–H stretch	2857	9	C–O stretch	1291
3	C=O stretch	1730–1738	10	C–O stretch	1240
4	C=O stretch	1715–1722	11	C–O stretch	1163
5	C=C stretch	1636	12	C–O stretch	1040
6	C–H bend	1466	13	C–H stretch	945
7	C–H bend	1412	14	C–H stretch	930

All shellac spectra show the characteristic peaks of this resin (Figure 8 and Table 4).

It is interesting to notice that it is not possible to differentiate between orange and bleached dewaxed shellac varnishes by FTIR (Figure 8), which is instead possible by UV imaging (Figures 2 and 7). These two techniques result hence to be complementary for the characterization of varnishes on copper alloys. However, different types of bleached dewaxed shellac appear similar under UV light in reason of the weak fluorescence of this material, especially in very thin layers (cf. 4.3). It is interesting to observe that, even though differences in hue are very hard to detect with naked eye, variations can be identified in the *L* a* b** colorimetric space. Nonetheless, caution must be applied in the interpretation of the results in case of weak fluorescence because even minor changes in the environmental condition during observation might affect the hue and lightness of the resulting color. These changes are on the contrary less noticeable when the fluorescence is strong and very tinted.

As by UV imaging, it is possible to detect by FTIR the presence of wax in the varnish, appearing as a characteristic double peak at 730 and 720 cm^{-1} [27] (p. 107), [28] (pp. 100–102); these peaks are indeed visible in the non-dewaxed orange varnish B (sample Smd_B) and in shellac wax (sample Sm_W), as well as, partially, in the manually dewaxed shellac varnish 8V (samples Sb_8V-a and Sb_8V-b), which clearly still contains traces of wax (Figure 9). Wax was instead more efficiently manually removed in varnishes 1L and 2L (samples Sb_1L-d and Sb_2L-d).

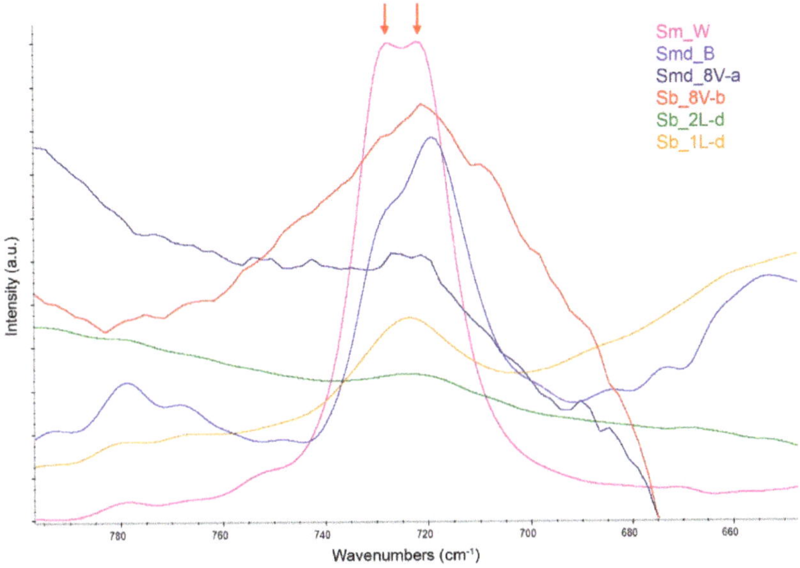

Figure 9. Characteristic peaks of shellac wax.

4.2. Comparison between Various Thicknesses of the Same Varnish

Figures 3 and 10 show that the thickness of the varnish layer affects the intensity of its fluorescence under UV light. Although the fluorescence of bleached dewaxed shellac is not always detectable with naked eye, it is possible to observe that thicker layers of varnish have a higher lightness value (L^*) and are more visible under UV light. Sample Mid_A-11, coated by industrial dip-coating with a varnish layer 11 μm thick, shows in fact the clearest fluorescence among all samples varnished with recipe A. Moreover, the fluorescence color is distinctly visible with naked eye and it is distinguishable from the more orange fluorescence of the raw shellac. On the other hand, it is difficult to discriminate between different types of support and different varnish application modes, especially for very thin layers of varnish. Looking at thicker layers, however, it seems that these two variables do not affect the intensity of the fluorescence, as shown by the similar $L^* a^* b^*$ coordinates of samples Mid_A-2 (mirror-polished substrate, industrial dip-coating) and Sb_1L-b (satin substrate, brush application), both having a varnish thickness of 2 to 3 μm. Nevertheless, it must be kept in mind that external factors might affect in a stronger way weaker fluorescence during observation, as already mentioned in Section 4.1.

Considering the orange dewaxed shellac, the sample coated with two layers of varnish 2L (Sb_2L-d) shows a stronger fluorescence than the sample having one layer (Sb_2L-b).

Concerning the infrared spectroscopy, the thickness of the analyzed layer seems to have an impact on the spectrum quality. Thicker samples lead in fact to the saturation of the C–H and C=O stretching peaks, and a higher thickness might be the cause of an anomalous series of waves deforming the baseline around 3100 cm^{-1} and between 2800 and 1800 cm^{-1} (Figure 11) [29]. More research needs however to be done to confirm this theory.

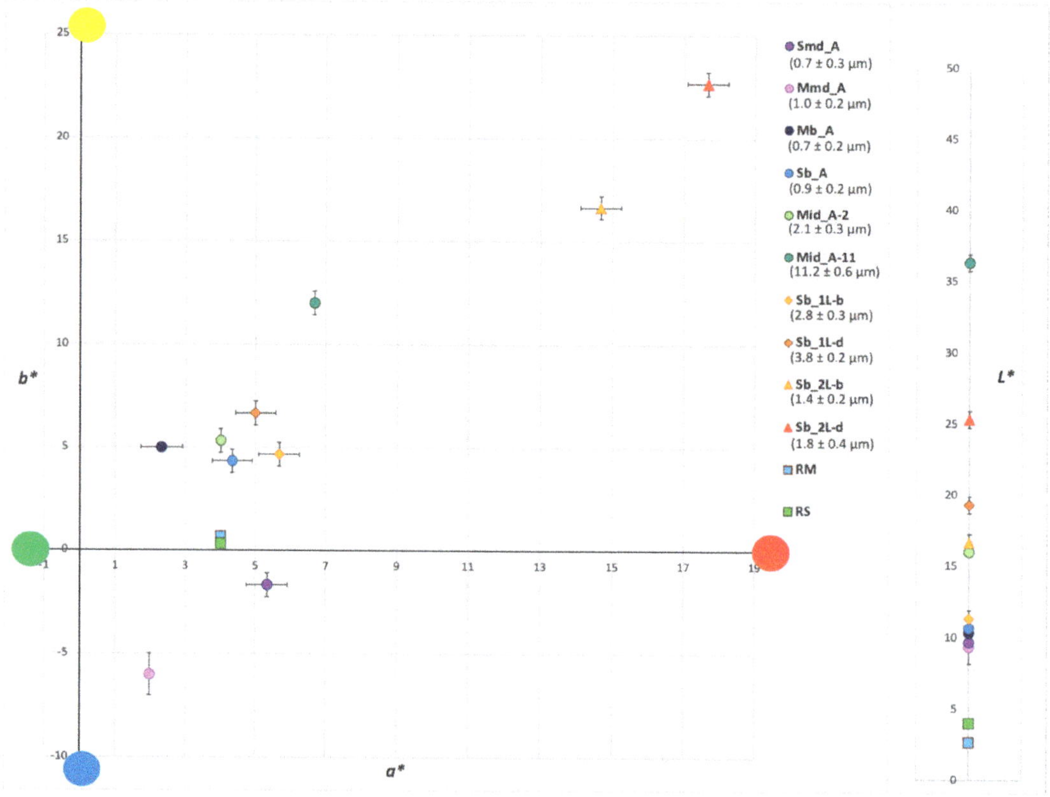

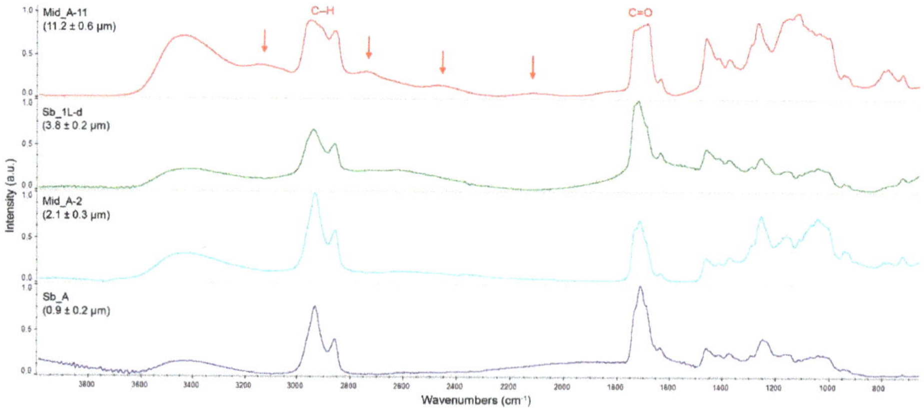

Figure 10. L^* a^* b^* colorimetric coordinates of the fluorescence under UV light of coatings with different thickness (cf. Figure 3). ●: varnish A; ◆: varnish 1L; ▲: varnish 2L; ■: non-varnished references.

Figure 11. Impact of the thickness on the FTIR spectrum quality. The arrows point to the "waves" indicated in the text.

4.3. Comparison of the Characteristics before and after Artificial Ageing

Plotting the *L* a* b** coordinates of the UV fluorescence of the samples before and after ageing (Figure 12, cf. Figure 4) it is possible to see that no change is noticeable for the industrially dewaxed bleached shellac samples with weak fluorescence (Smd_1L-ast and Smd_1L-pla), whereas a slight increase in lightness is found for the thicker and more fluorescent sample Smd-AS. This might be due to the fact that slight changes in hue and lightness are more difficult to detect in case of low fluorescence. On the other hand, samples coated with the manually dewaxed varnishes 1L, 2L, and 8V show a shift towards bluish-greenish hues with ageing, with the exception of the orange shellac applied in two layers (sample Sb_2L-c/-d). The difference is higher for less fluorescent bleached shellac varnish 1L, with ΔE values of 12.8 (Sb_1L-c/-d) and 7.1 (Sb_1L-a/-b), and lower for the more fluorescent unbleached shellac varnishes 2L and 8V, with ΔE values of 4.7 (Sb_2L-a/-b) and 3.6 (Sb_8V-a). It needs to be noted that, as explained in Section 3.2, it was not possible to compare the same sample before and after ageing for varnishes 1L and 2L, which might partially affect the difference in hue of these samples. More research needs to be done on a larger set of samples in order to confirm these preliminary results.

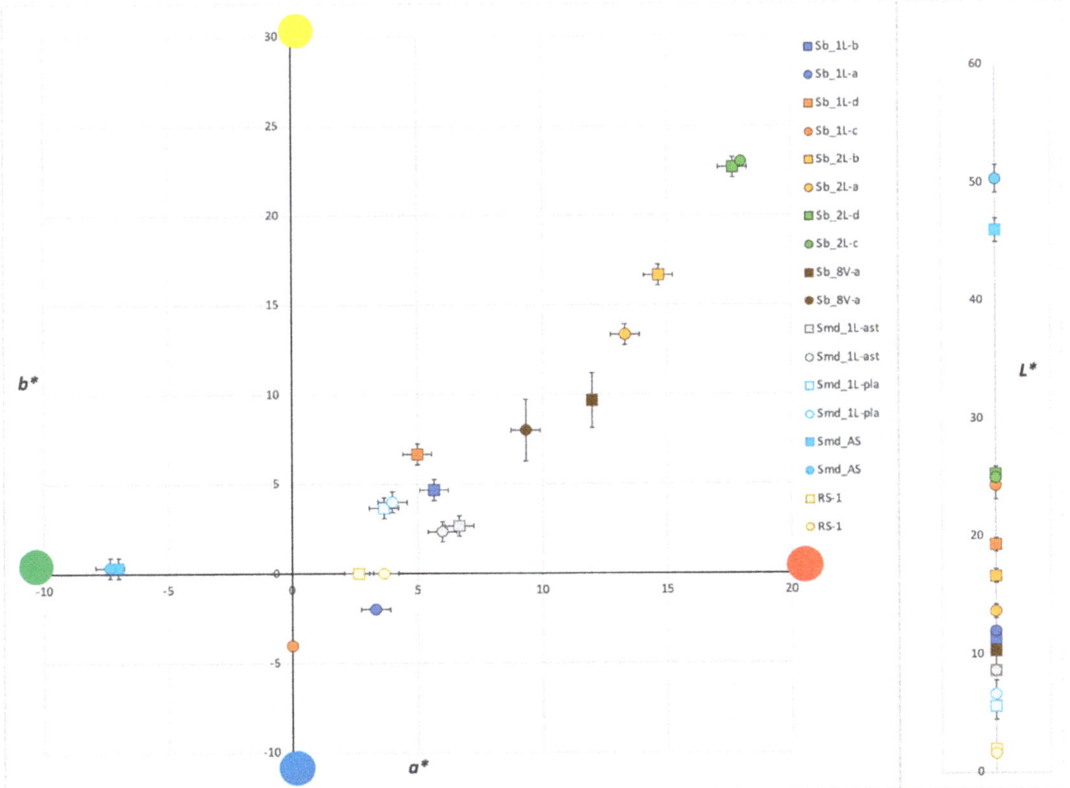

Figure 12. *L* a* b** colorimetric coordinates of the varnishes' fluorescence under UV light before (■) and after (●) ageing (cf. Figure 4).

Results of the FTIR analysis, on the other hand, show no difference between the spectra of the samples before and after ageing (Figure 6).

5. Conclusions

The UV imaging and FTIR analysis of mock-up varnished brass coupons allowed the assessment of the effectiveness of these techniques in the identification of different grades of shellac. The observation under UV light of the samples resulted to be an efficient method to discriminate between bleached dewaxed shellac and raw dewaxed shellac; however, the limit of detection of the UV fluorescence of the varnish relates to the thickness of the coating, especially for the more refined resin. For thicknesses ≤ 1 µm it is in fact very difficult to detect the presence of bleached dewaxed shellac with naked eye under UV light, although a slight change in the L^* a^* b^* coordinates does occur. This shows the need to use a particularly strong radiation source when it comes to observing objects in situ in order to be able to detect even the thinnest layers. FTIR spectroscopy, on the other hand, proved to be effective in the detection of very thin layers of shellac-based varnishes. Besides, a greater thickness of the coating seems to affect the IR spectrum quality, leading to saturation and deformation of the IR bands. This effect was observed in the present study for varnishes thicker than 10 µm, but further investigation is necessary to verify the behavior between 2 and 10 µm. Contrary of the UV imaging, it is not possible to discriminate between different grades of shellac by FTIR spectroscopy. These two techniques proved therefore to be complementary in the detection and identification of shellac varnishes on copper alloys.

From a qualitative point of view, bleached dewaxed shellac fluorescence tends towards neutral colors, whereas orange shellac and seedlac fluorescence exhibits a more orange hue. The presence of additives in the varnish can heavily affect the color of the sample under UV light, as showed by the ready-to-use "Astra" shellac containing traces of rosin and exhibiting a greener fluorescence. Moreover, the presence of shellac wax, easily detectable by FTIR spectroscopy, increases by far the fluorescence of the varnish. It needs to be noted that, when analyzing real artefacts, we focused on the examination of portions of the objects presenting a varnish in a good condition; therefore, no investigation was carried out on the mock-up samples on the possible influence of metal corrosion on the varnish fluorescence under UV light, as well as on the interaction between the varnish and the metal. These factors could further affect the intensity and the color of the varnish fluorescence.

No changes were observed in the FTIR spectra of samples before and after artificial ageing, nor in the UV fluorescence of the majority of the coupons. Some of the varnishes showed a shift towards bluish-greenish hues with ageing but it is not clear how other factors might have affected this result. More research needs to be done on a larger set of samples in order to confirm this result.

Author Contributions: Conceptualization, methodology and validation: M.T.G., L.B. and J.S.; formal analysis and investigation: M.T.G. and L.B.; writing—original draft preparation: M.T.G.; writing—review and editing: M.T.G., L.B. and J.S.; visualization: M.T.G.; supervision: L.B. and J.S.; project administration and funding acquisition: J.S. All authors have read and agreed to the published version of the manuscript.

Funding: This research was funded by the Haute Ecole Spécialisée de Suisse Occidentale, Grant No. 100779/DAV-RAD19-06, and co-funded by Haute Ecole Arc conservation-restauration.

Institutional Review Board Statement: Not applicable.

Informed Consent Statement: Not applicable.

Data Availability Statement: The data presented in this study are available within the article and in [5].

Acknowledgments: The authors would like to thank Laure Jeandupeux, (Haute Ecole Arc, Neuchâtel, Switzerland, medical devices), David Grange (HES professor, Haute Ecole Arc, Neuchâtel, Switzerland), and Stephan Ramseyer (Haute Ecole Arc, Neuchâtel, Switzerland, surface engineering). We would also like to thank the Lathema laboratory of University of Neuchâtel for the use of the FTIR with external module.

Conflicts of Interest: The authors declare no conflict of interest. The funders had no role in the design of the study; in the collection, analyses, or interpretation of data; in the writing of the manuscript, or in the decision to publish the results.

References

1. Halleux, R. *Les Alchimistes grecs. Tome I: Papyrus de Leyde—Papyrus de Stockholm—Recettes*, 3rd ed.; Les belles lettres: Paris, France, 2019.
2. Verlet, P. *Les bronzes dorés francais du XVIIIe siècle*; Editions A & J Picard: Paris, France, 1987.
3. Mounier, A.; Daniel, F.; Bechtel, F. L'illusion de l'or. *ArcheoSciences* **2009**, *33*, 397–403. [CrossRef]
4. Perego, F. *Dictionnaire des matériaux du peintre*; Éditions Belin: Paris, France, 2005.
5. Schröter, J.; Michel, A.; Mirabaud, S.; Brambilla, L.; Paris, C.; Bellot-Gurlet, L. Transparent varnishes on copper alloys dating from the 19th century: Characterization and identification strategies. In Proceedings of the interim meeting of the ICOM-CC metals working group, Neuchâtel, Switzerland, 2–6 September 2019.
6. Buzzegoli, E.; Keller, D. Ultraviolet fluorescence imaging. In *Scientific Examination for the Investigation of Paintings: A Handbook for Conservator-Restorers*; Pinna, D., Galeotti, M., Mazzeo, R., Eds.; Centro Di: Florence, Italy, 2009; pp. 204–206.
7. Thackray, A. A methodology for the conservation of furniture mounts. *V A Conserv. J.* **2014**, 62.
8. Cosentino, A. Practical notes on ultraviolet technical photography for art examination. *Conserv. Património* **2015**, *21*, 53–62. [CrossRef]
9. Measday, D. A summary of ultra-violet fluorescent materials relevant to conservation. *AICCM Natl. Newsl.* **2017**, 137. Available online: https://bit.ly/3pjAq8U (accessed on 10 February 2021).
10. Umney, N.; Rivers, S. *Conservation of Furniture*; Routledge: New York, NY, USA, 2013.
11. The Conservation and Art Materials Encyclopedia Online (CAMEO). Available online: http://cameo.mfa.org/wiki/Category:Materials_database (accessed on 10 February 2021).
12. Thomson, C. Last but not least, examination and interpretation of coatings on brass hardware. In Proceedings of the AIC Annual Meeting, Albuquerque, NM, USA, 3–8 June 1991; pp. 1–15.
13. Long, D. The treatment of false gilding: A case study. In *Gilded Metals: History, Technology and Conservation*; Drayman-Weisser, T., Ed.; Archetype Publications Ltd.: London, UK, 2000; pp. 319–327.
14. Jeanneret, R. Approche Pluridisciplinaire Pour le Traitement de Conservation-Restauration d'un Quart de Cercle Mural du 18ème Siècle du Musée des Confluences de Lyon. Master's Thesis, HES-SO, Neuchâtel, Switzerland, 2010.
15. Galeotti, M.; Joseph, E.; Mazzeo, R.; Prati, S. Transform infrared spectroscopy. In *Scientific Examination for the Investigation of Paintings: A Handbook for Conservator-Restorers*; Pinna, D., Galeotti, M., Mazzeo, R., Eds.; Centro Di: Florence, Italy, 2009; pp. 151–156.
16. Echard, J.-P. Le vernis des instruments de musique: Principe et spécificités. In *Art et Chimie: Les Polymers*; CNRS Editions: Paris, France, 2002; pp. 75–80.
17. Lanterna, G.; Giatti, A. Caratterizzazione non invasiva delle vernici da ottone degli strumenti scientifici: Ricette storiche, realizzazione di provini verniciati, ricerca analitica e applicazioni "in situ" su strumenti storici. *OPD Restauro* **2014**, *26*, 165–180.
18. Salvadori, B.; Galeotti, M.; Porcinai, S.; Cagnini, A. Non-invasive FT-IR reflection to assess and monitor coatings on metals, In proceedings of the New strategies for diagnostics of conservation treatments conference, Amsterdam, The Netherlands, 7–8 February 2019.
19. Tripier-Deveaux, A.-M. *Traité Théorique et Pratique sur l'art de Faire les Vernis: Suivi de Deux Mémoires, l'un sur les Dangers Qui menacent les Peintures Vernies d'extérieurs, l'autre sur les Précautions à Prendre Pour Assurer aux Revernissages la Même Durée Qu'aux Vernissages Faits sur les Peintures Fraîches*; Librairie scientifique-industrielle de L. Mathias: Paris, France, 1845.
20. Société de l'École d'horlogerie de Paris; Chambre syndicale de l'horlogerie de Paris. *Revue Chronométrique*; Organe des Sociétés D'horlogerie et des Chambres Syndicales: Paris, France, 1876.
21. Roseleur, A. *Guide Pratique du Doreur, de L'argenteur et du Galvanoplaste: Manipulations Hydroplastiques*; Plazanet: Paris, France, 1873.
22. Rie (de la), E.R. Fluorescence of paint and varnish layers (Part I). *Stud. Conserv.* **1982**, *27*, 1–7. [CrossRef]
23. Mills, J.S.; White, R. *The Organic Chemistry of Museum Objects*; Butterworth & Co.: London, UK, 1987.
24. ISO 16474-3:2013. *Paints and Varnishes—Methods of Exposure to Laboratory Light Sources—Part 3: Fluorescent UV Lamps*; ISO: Geneva, Switzerland, 2013.
25. Simpson-Grant, M. The Use of Ultraviolet Induced Visible-Fluorescence in the Examination of Museum Objects, Part 2. *Conserve O Gram* **2000**, *1*, 1–4.
26. The Conservation and Art Materials Encyclopedia Online (CAMEO). Available online: http://cameo.mfa.org/wiki/Rosin (accessed on 10 February 2021).
27. Derrick, M.R.; Stulik, D.C.; Landry, J.M. *Infrared Spectroscopy in Conservation Science*; Getty Conservation Institute: Los Angeles, CA, USA, 1999.
28. Derry, J. Investigating Shellac: Documenting the Process, Defining the Product. A Study on the Processing Methods of Shellac, and the Analysis of Selected Physical and Chemical Characteristics. Master's Thesis, University of Oslo, Oslo, Norway, 2012.
29. Nakata, Y.; Murakami, S. *FTIR Talk Letter*; Shimadzu: Kyoto, Japan, 2010; Volume 15.

Article

Drying Oil and Natural Varnishes in Paintings: A Competition in the Metal Soap Formation

Tommaso Poli [1,*], **Oscar Chiantore** [1], **Eliano Diana** [1] **and Anna Piccirillo** [2]

[1]   Dipartimento di Chimica, Università degli Studi di Torino, Via Pietro Giuria 7, 10125 Torino, Italy; oscar.chiantore@unito.it (O.C.); eliano.diana@unito.it (E.D.)

[2]   Centro Conservazione e Restauro "La Venaria Reale", Via XX Settembre 18, 10078 Venaria Reale, Italy; anna.piccirillo@centrorestaurovenaria.it

*   Correspondence: tommaso.poli@unito.it

Abstract: Metal soaps formation is a well-known issue in oil paintings. Along the lifetime of the painting, carboxylic acids coming from drying oil (free fatty acids, acids from hydrolysis of triglycerides and from oxidation processes) can react with cations of some pigments (in particular, smalt, lead white and zinc white) forming the related carboxylic salts. As observed by many authors, the formation of these carboxylates, with the tendency to migrate and to aggregate, not only modifies the behavior and the aspect of the paint film but also complicates the cleaning approach. In previous works we have demonstrated that a similar pigment reactivity is possible even in presence of natural resins (such as colophony, dammar, mastic, etc) historically used as final varnishes on paintings. In this case, in the reactions the terpenic acids, among the main components of the resins, are involved. In this work, the carboxylates formation kinetics has been studied starting from two representative acids (palmitic and abietic) of painting oils and natural varnishes. Successively, the reactivity of the palmitic acid with the potassium abietate and of the abietic acid with the potassium palmitate has been verified. This investigation aims at clarifying in which way terpenic acids can be involved in the metal soaps reactivity confirming that also surface varnishes may play a significant role in the carboxylates formation and reactivity. It is important to keep in mind that a finishing varnish can be removed and reapplied many times during the lifetime of a painting, thus renewing the provision of reactive terpenic acids at the interface of the painted layers.

Keywords: paintings; metal soaps; natural resins; varnishes; smalt

Citation: Poli, T.; Chiantore, O.; Diana, E.; Piccirillo, A. Drying Oil and Natural Varnishes in Paintings: A Competition in the Metal Soap Formation. *Coatings* **2021**, *11*, 171. https://doi.org/10.3390/coatings11020171

Academic Editor: Andréa Kalendová

Received: 18 December 2020

Accepted: 28 January 2021

Published: 31 January 2021

Publisher's Note: MDPI stays neutral with regard to jurisdictional claims in published maps and institutional affiliations.

1. Introduction

One of the most common decay pattern of oil paintings is the formation, along the lifetime of the work of art, of metal soaps [1–6]. Carboxylic acids present in the drying oil can react with cations of some pigments (in particular, smalt, lead white and zinc white) forming the related carboxylic salts. Few free fatty acids are present in the fresh oil but during the ageing many factors such as the hydrolysis of triglycerides can increase drastically the concentration of free acids sites. Moreover, the characteristic of drying oils, linseed oil in particular, is the presence of double bonds that allow the crosslinking process. These reactive sites are also prone to oxidative processes that easily lead to the chain breaking and the formation of new acid sites [7–12]. As observed by many authors, these carboxylates, defined metal soaps, with the tendency to migrate in the painted layer and to aggregate, not only modify the chemical-physical properties but also the aesthetic of the painting by generating efflorescences and in the worst cases protrusions [13–17]. Many studies have been carried out in order to understand the chemical structure of these aggregates and in which way the aggregation takes place. Main results suggest that cations migrate from the pigment surface to the bulk of the binder (the drying oil) shifting from an acid site to another. The polar mobile molecules aggregate in three-dimensional structures, a polymeric/ionomeric network, exploiting bifunctional "bricks" such as the azelaic acids

123

molecules coming from the oxidation of the unsaturated fatty acids [18]. The presence on surface of these protrusions and in general of metal soaps make the cleaning approach to the painting definitely complicate and delicate. Metal soaps cannot be easily removed from painting layers and surfaces since they are poorly soluble in traditional solvents used in conservation and, moreover, not all the conservators agree on their removal.

In previous works authors demonstrated that a similar pigment particles reactivity is possible even in presence of natural resins (such as colophony, dammar, mastic, etc) historically used as final varnishes on paintings [19–24]. The reaction involves the terpenic acids (mono, di, and triterpenic), the main components of this kind of resins. In a theoretical situation, being "protected" by the binder, very few pigment can come in contact with the natural resins (historical recipes of varnishes are always a mixture of resins) since they are applied as final superficial varnish. It is important to point out that, in the past, artists often used to add some resin in the drying oil recipes and the first varnish application occurs early when the oil is not perfectly dried [25]. Moreover, natural varnishes were applied through the use of a common solvent (the turpentine): factors that can favor the interaction of the two phases. Moreover, the usual conservation treatments of a painting include the varnish substitution [26]: new fresh terpenic varnish in solvent is applied and come in contact with painted layers depleted in the organic component (the binder) as a consequence of previous usual solvent-based removal of the old varnish. It must be considered that this operation could have taken place several times in the life of a painting.

In this work the carboxylates formation and kinetics have been studied by Fourier Transform Infrared (FTIR) spectroscopy, working in transmission and with an Attenuated Total Reflectance (ATR) accessory. Mid-infrared spectroscopy has been proven to be one of the most effective and sensitive techniques for the detection of metal soaps. They are characterized by a typical signal between 1520 and 1590 cm^{-1} depending on the cation involved [1,5,6,9,10,12–16,19,21–23]. exploiting two representative acids of painting oils and natural varnishes (palmitic and abietic acid respectively) and potassium hydroxide (KOH) as cation source. The potassium cation has been chosen since it is involved in an important degradation reaction in the field of painting conservation: the discoloration of smalt (see below) [27–32]. Smalt is an historically widely employed blue pigment constituted by a fine grinded potassium glass and it is well-known that is one of the most reactive pigment in presence of drying oil. This simplified model system was useful for interpreting the results about the reactivity and kinetics, under the same laboratory conditions, of the mixture made with drying oil, natural resins and smalt. Successively, the reactivity of the two acids has been investigated by putting them in contact with the already formed potassium metal soap: palmitic acid with potassium abietate and abietic acid with potassium palmitate. The whole work was aimed at clarifying in which way terpenic acids can be involved in the metal soaps reactivity, and at confirming that also surface varnishes may play a significant role in the carboxylates formation and reactivity.

2. Materials and Methods

Palmitic acid (Hexadecanoic acid, ≥99%), Abietic acid (technical, ~75%) and potassium hydroxide (pellets for analysis EMSURE®) have been supplied by Sigma-Aldrich (Saint Louis, MO, USA.). Linseed stand oil (ref. 73,200) and smalt (ref. 10,000) have been supplied by Kremer Pigmente (average grinding of smalt around 10 microns).

Silicon float zone polished windows (13 mm diameter, by 1mm thickness) have been supplied by Crystran Ltd. (Dorset, UK).

FT-IR transmission spectra (64 scans) have been carried out on a Bruker Vertex 70 spectrophotometer (Bruker, Billerica, MA, USA.), working in the spectral range from 4000 to 400 cm^{-1} with an average spectral resolution of 4 cm^{-1}.

Mixtures have been applied on the silicon wafer with a plastic spatula (in order to avoid an eventual surface scratching) in a thin layer that allows a good transmission signal with absorbance of the strongest band around 1. Once applied the mixture the recording of spectra started, and a spectrum every 15 min has been recorded for a total of 700 min.

Successively, a spectrum every hour has been recorded till 240 h. The kinetic measurements of abietate salts mixed with palmitic acid and of palmitate salts mixed with abietic acid have been carried on a single angle ATR equipment (Harrick, Pleasantville, NJ, USA.) in order to avoid saturation problems (we have the necessity to add material). To prepare the abietate and the palmitate for the reaction with the other acid, the stoichiometric ratio acid/KOH has been calculated and the amount of KOH has been reduced by 10% in order to be sure that there was no KOH unreacted and waited 20 days mixing twice a day. The amount of acid successively added, in order to verify the possibility of cation exchange, is calculated to have almost (the purity of commercially available abietic acid is ~75%) the same number of acid sites.

3. Results and Discussion

Since XVIIth century smalt is a widely used blue pigment. It is a potassium glass with color coordination centers occupied by cobalt cations. During the lifetime of the painting, humidity and acids force the leeching of cations from the glass structure favoring the carboxylates formation. Bibliography [27–32] suggests that metal soaps in smalt painted layers are mostly constituted by palmitic acid and potassium extracted from smalt pigment. The reaction can be formalized in this way where R–COOH is a generical fatty acid of the linseed stand oil:

$$\equiv Si\text{-}O^- \; K^+ + R\text{-}COO^- \; H_3O^+ \rightarrow \equiv Si\text{-}OH + R\text{-}COO^- \; K^+ + H_2O$$

This is a very important reaction in the field of conservation since is the main cause of the discoloration of smalt in ancient paintings. The first step of this work aimed at verifying the reactivity and the kinetics of a stand oil/smalt system. These data are important as reference in order to verify if the simplified modeling of the successive steps is sensible. The behaviour of a mixture of 1:1 (w/w) stand oil/smalt applied on a silicon wafer is therefore compared with the behaviour of a mixture of linseed oil potassium hydroxide (KOH) and palmitic acid with KOH.

In Figure 1 it is possible to observe the growth of a broad band ranging from 1525 to 1600 cm^{-1}, centered at 1572 cm^{-1}, related to the formation of metal soap in the experimental condition (fine grinding of smalt around 10 microns, 22 °C, 45% RH% and applied with a spatula on the silicon wafer) and the corresponding increase of the OH broad band around 3400 cm^{-1} related to the formation of water molecules. The carbonyl of carboxylic salts bands appears at lower wavenumbers (from ~1700 to ~1550 cm^{-1}) than the carbonyl of the acid in free form or of the esters (see Table 1). In Table 1, the main signals, in the mid-infrared spectral range, of linseed oil, colophony, abietic and palmitic acid, are reported according to literature [6,8,22,23,33–36]. Many factors can contribute to the broadening of the band even in the experimental conditions; this broadening is related to the presence of different kind of fatty acids (mono and di-functional), the carboxylates partially coordinated around the pigment grains, the free ones and the aggregated ones. The experimental conditions, in particular the fine grinding, aimed, unsuccessfully, to obtain a sharper band of carboxylates. In the real conditions this band is even broader due to the increased complexity of the system induced by the presence of other pigments and the oxidative ageing and crosslinking of the oil. A broad band would not permit to distinguish metal soaps coming from oil or resin acids since they absorb in a range very close as reported in literature [22,23].

$$KOH + R\text{-}COO^- \; H_3O^+ \rightarrow R\text{-}COO^- \; K^+ + H_2O$$

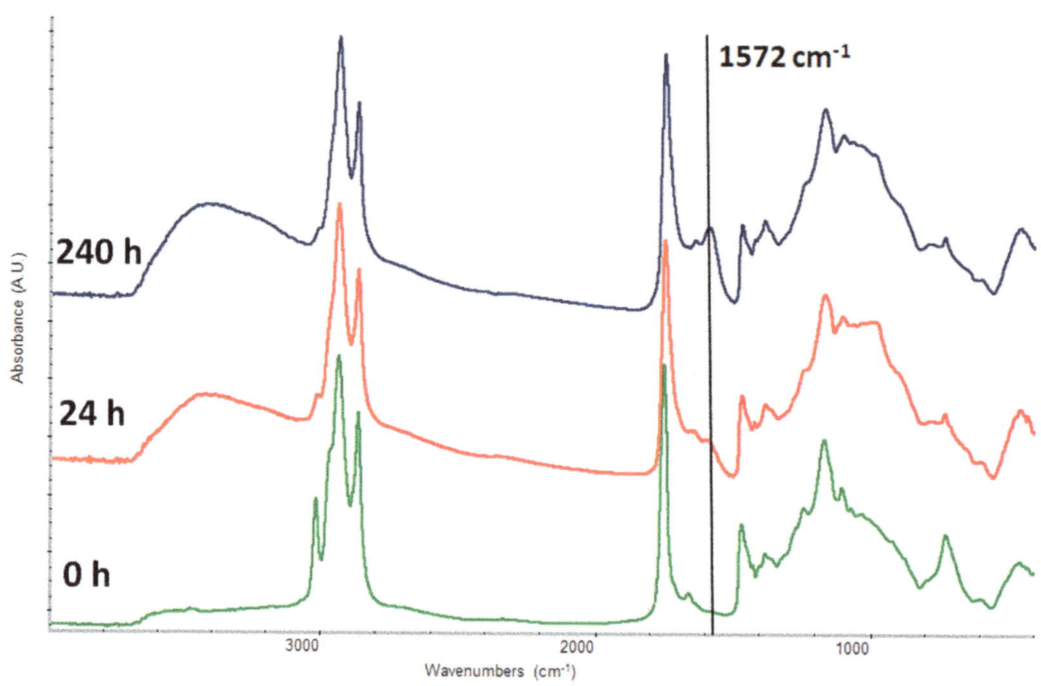

Figure 1. Kinetics of smalt carboxylate formation from a mixture 1:1 (w/w) of stand oil and smalt. The black arrow shows the carboxylate band.

Table 1. Table reports the position (in wavenumbers, cm^{-1}) and the assignments of the main bands of linseed oil, colophony, palmitic acid and abietic acid.

Assignment	Linseeed Oil	Palmitic Acid	Colophony	Abietic Acid
Stretching OH	3465	-	3420	3422
C=C lin insaturation	3011	-	-	-
CH_3 asym. stretching	-	2956	-	-
CH_2 asym. stretching	2924	2916	2928	2928
CH_2 sym. stretching	2852	2849	2865	2865
C=O carbonyl stretching esters	1740	-	-	-
C=O carbonyl stretching acids	-	1694	1710	1689
C=C conjugated with C=C or with C=O	1643	-	1622	1622
C=O carbonyl stretching K salts	1572	1564	1564	1544
CH_2 asym. and sym. bending	1469	-	1457	1465
CO stretching + OH vib	1423	-	-	1440
CH_3 asym. and sym. bending	1385	-	1381	1393
OH bending	-	-	-	-
CCO bending	1241	1315	1241	-
COC asym stretching	1169	1299	1161	1280

Table 1. *Cont.*

Assignment	Linseeed Oil	Palmitic Acid	Colophony	Abietic Acid
CH$_2$ and C–C crystalline progression (wag. and twist.)	-	1290–1180	-	-
C–O asym. stretching	1110	-	-	-
C–O sym. stretching	-	-	1038	-
C–C ring	-	-	983	1199
C–C ring	-	-	899	1161
C–C	979	-	-	962

As expected, substituting the smalt with KOH the reaction proceeds faster (Figure 2) according to the above reported reaction. The ratio surface/volume is by far more favorable, and the potassium is immediately available as basic hydroxide. Nevertheless, while the reaction proceeds the sharp band at 1564 cm^{-1} starts broadening. The slight shifting of the maximum from 1572 to 1564 cm^{-1} and the broadening confirms the fact that we are observing" different" potassium carboxylates: the neo-formed and, over time, the ones involved in more complex interactions with triglycerides and fatty acids (some of them partially cross-linked as pointed out by the decreasing of the double bond signal at 3011 cm^{-1}) that re-organize in new aggregated ionomeric systems [1,9,13,15]. The double bonds are present in the unsaturated fatty acids of linseed oil and they are the responsible of the drying process.

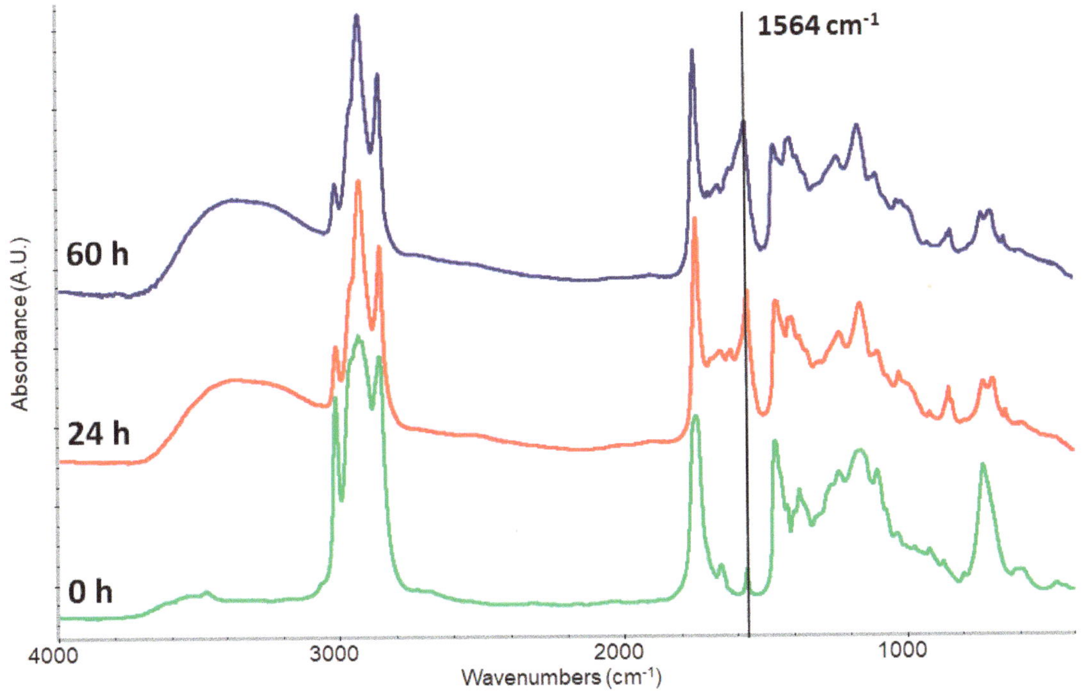

Figure 2. Kinetics of potassium carboxylate formation from a mixture 1:1 (*w/w*) of stand oil and KOH. The black arrow shows the carboxylate band.

In order to obtain a definitely sharp band of K carboxylate we need to further simplify the system by substituting the complex natural mix of molecules inside the drying oil with a single representative molecule: palmitic acid. In Figure 3 we may observe that the carboxylate band, centered at 1564 cm^{-1}, remains sharp and perfectly resolved.

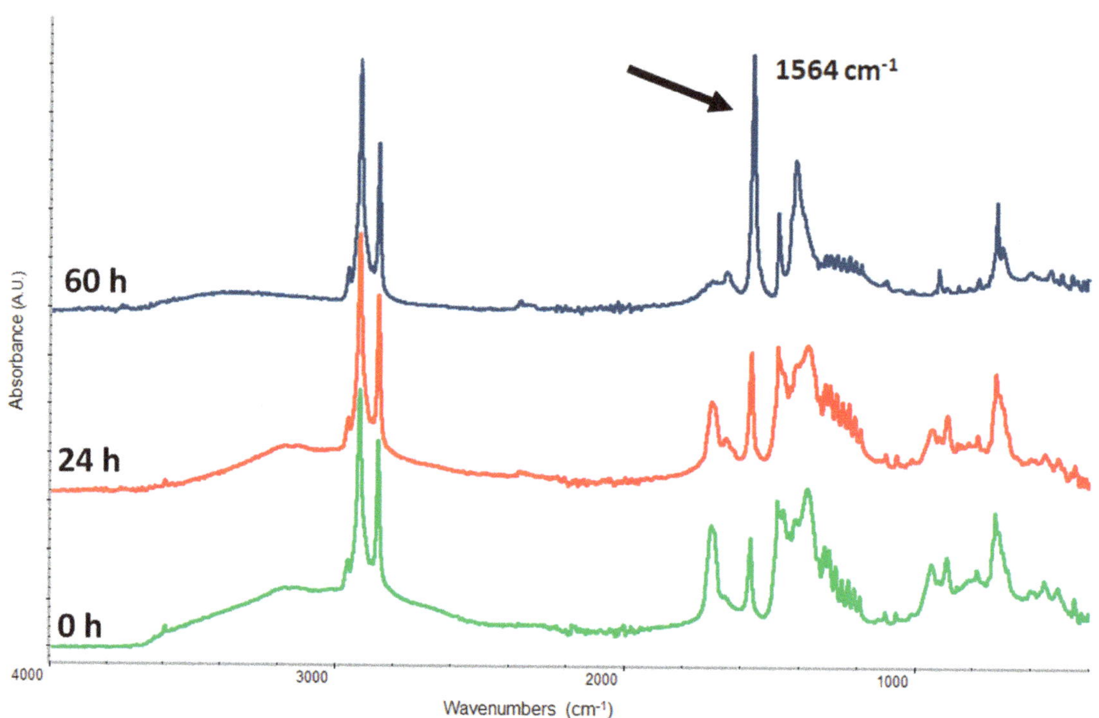

Figure 3. Kinetics of potassium carboxylate formation from a mixture 1:1 (*w/w*) of palmitic acid and KOH. The black arrow shows the carboxylate band centered at 1564 cm^{-1}.

Even in this case, where the binder is the natural resin, in order to obtain an enough resolved band of carboxylates (see Figure 4) it has not been possible to use a terpenic resin (colophony or dammar) which would form a broad band as assessed in a previous paper, but we used a representative molecule: abietic acid. Abietic acid has been chosen since is one of the few terpenic acids, characteristic of a natural resin, available on the market. Our measurements confirm that even the terpenic resins, the historical varnishes, on paintings can contribute to the formation of metal soaps and then to the smalt decay processes. The certainty of obtaining well resolved spike bands makes it possible the second part of this investigation where the reactivity of palmitic acid with an already formed potassium abietate, and vice versa, have been assessed. Here we want to verify the possibility of cation exchanging between drying oil metal soaps and the acid coming from a terpenic resin. In Figure 5 it is shown what happens when palmitic acid and abietic acid are mixed with an excess of potassium hydroxide. The expected reaction is reported below where P–COOH is the palmitic acid and A–COOH the abietic:

$$2KOH + P\text{-}COO^- \ H_3O^+ + A\text{-}COO^- \ H_3O^+ \rightarrow P\text{-}COO^- \ K^+ + A\text{-}COO^- \ K^+ + 2H_2O$$

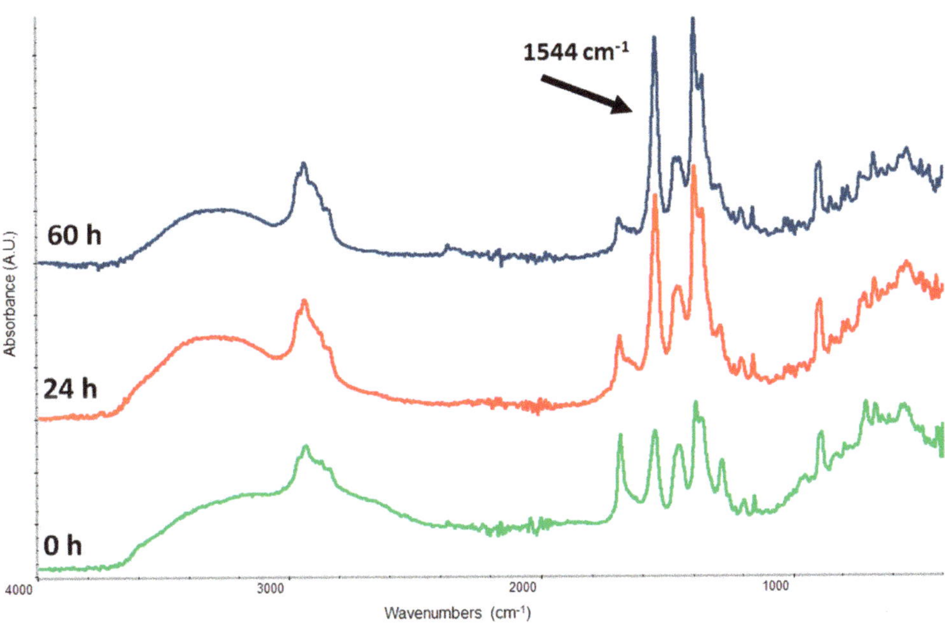

Figure 4. Kinetics of potassium carboxylate formation from a mixture 1:1 (w/w) of abietic acid and KOH. The black arrow shows the carboxylate band centered at 1544 cm^{-1}.

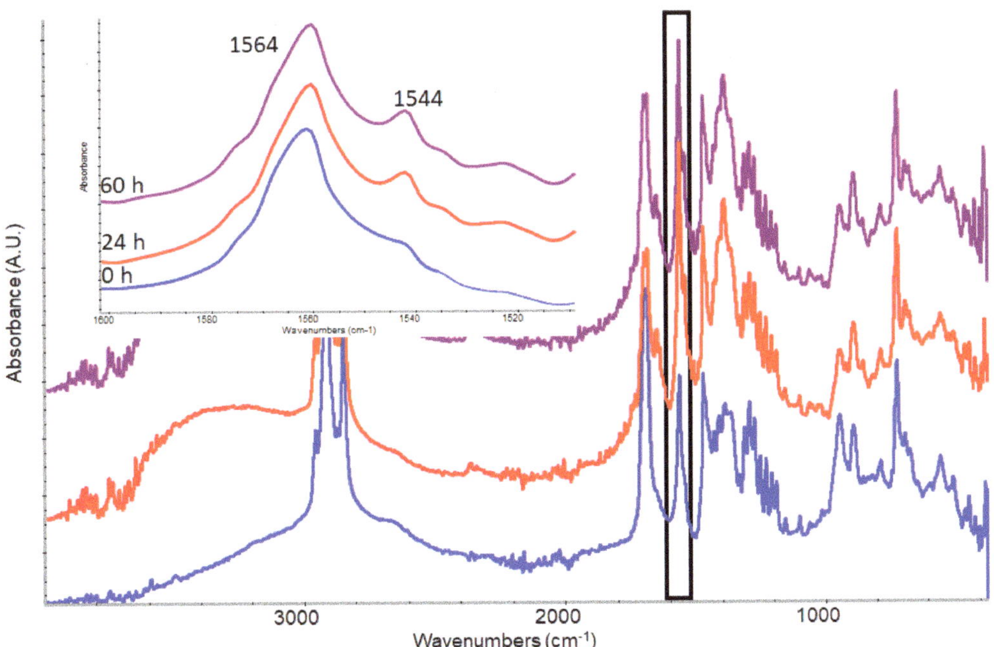

Figure 5. Kinetics of potassium carboxylates formation from a mixture 1:2:1 (w/w) of palmitic acid, KOH and abietic acid. In the box the spectral range from 1500 to 1600 cm^{-1} is highlighted.

The simultaneous formation of the two carboxylates occurs at the previously verified wavenumbers of 1564 cm^{-1} for the palmitate and 1544 cm^{-1} for the abietate. This means that when KOH is available both acids form the related carboxylate.

When palmitic acid is added to potassium abietate (prepared with a defect of KOH in order to be sure that no KOH was available) no palmitate formation is observed (Figure 6).

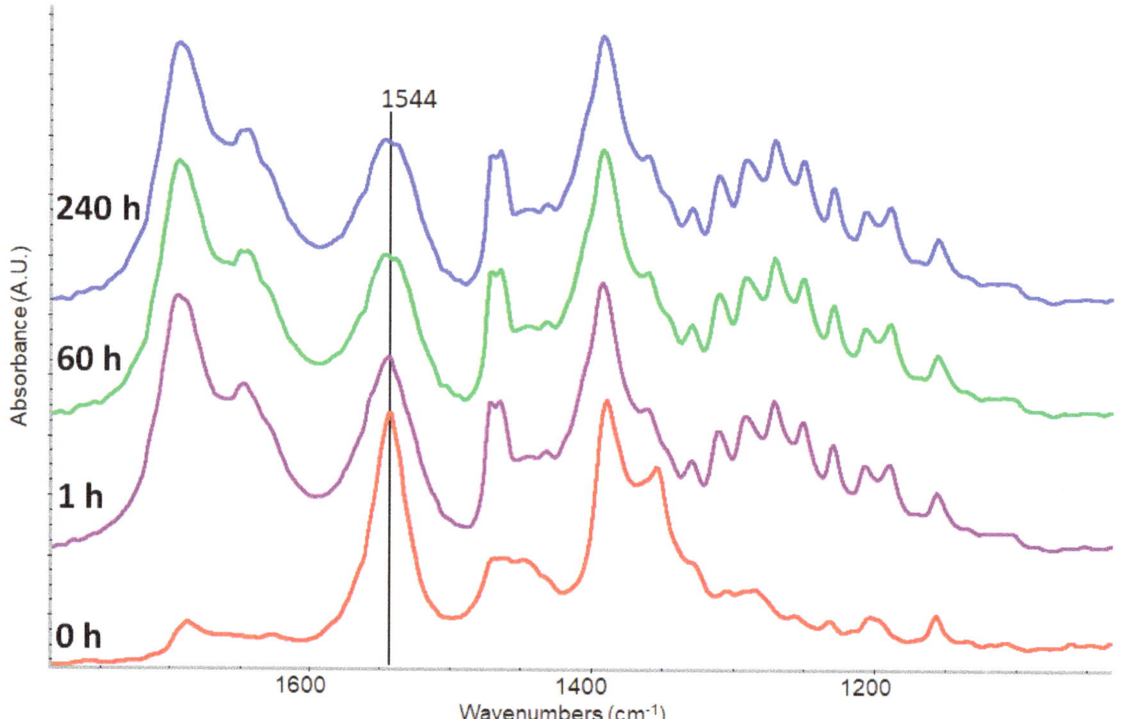

Figure 6. Kinetics of potassium abietate mixed with fresh added palmitic acid in the spectral range from 1100 to 1800 cm^{-1}. Before the mixing (red line), after 1 h (violet line), after 60 h (green line) and after 240 h (blue line).

The palmitic acid does not seem able to steal the cation of the salt based on abietic acid or at least not in a significant amount. After 60 h (green line) it may be observed only a splitting of the maximum probably related to the perturbation of COO–Me$^+$ bond by the acid function of the palmitic acid.

The situation is clearly different when abietic acid is added to an already formed potassium palmitate (Figure 7). After 1 h a well-shaped band at 1544 cm^{-1} appears and after 60 h the band of palmitate almost disappear: the abietic acid "snatches" the cation to palmitate as reported in the reaction below.

$$P\text{-}COO^- \ K^+ + A\text{-}COOH \rightarrow +P\text{-}COOH + A\text{-}COO^- \ K^+$$

After 240 h, the signal of palmitate is practically undetectable while the band of abietate (centered at 1544 cm^{-1}) becomes the main absorption in the critical spectral range. It's important to point out that all these reactions occur in solid, semi-solid phase trough an intimate but simple mixing and a poorer yield is expected. This behavior is very peculiar since pKa of abietic acid should be very similar to the palmitic acid one or slightly higher due to the methyl in α position. Further investigations are going on, taking in account the solubility of the formed salts vs. acid strength in semi-anhydrous conditions.

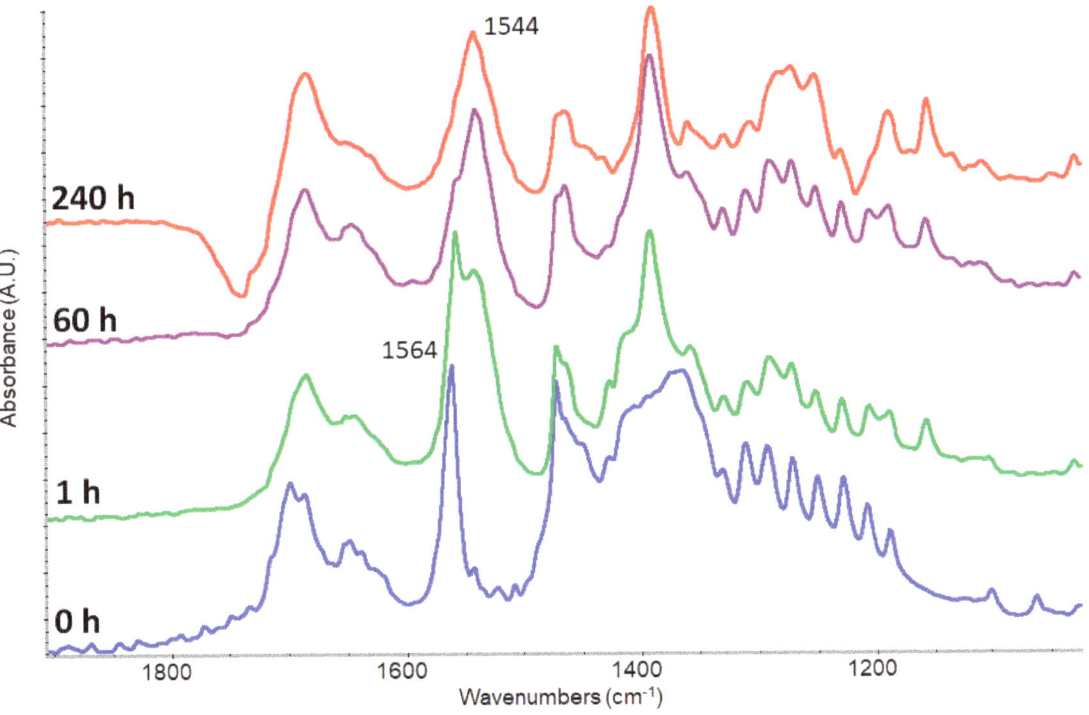

Figure 7. Kinetics of potassium palmitate mixed with fresh added abietic acid in the spectral range from 1100 to 1800 cm^{-1}. Before the mixing (blue line), after 1 h (green line), after 60 h (violet line) and after 240 h (red line).

This is an interesting result that may change the approach to the metal soaps formation on the painting surfaces. Natural resins not only can form metal soaps interacting with pigments as already previously assessed, but it is shown here that they can also successfully compete and "snatch" the cations from already formed fatty acid carboxylates. From the conservation point of view precise consequences immediately arise. Natural resins are mostly used as varnishes for surface finishing and can be reapplied many times in the life of a painting; fresh and reactive new resin come therefore in contact with already formed and aggregated soaps giving a contribution modifying and influencing the ionomeric structure growth. Ionomeric structures and metal soaps protrusions do not simply grow under the varnish but probably the varnish is deeply involved in the process.

4. Conclusions

The results here presented demonstrate that in the formation and growth of metal soaps the role of natural resins, used as varnishes or co-binder in oleo-resinous media, could be far more important than previously expected. Natural resins in fact are active (though underrated) players in the metal soap formation, growth, and aggregation through the reactivity of their acid constituents. The work confirms that the acid components of natural resins not only can provide "bricks" for the metal soap aggregation reacting with pigments but also can change the composition of already formed aggregates subtracting cations and freeing acids functions of fatty acids. This possibility has been verified with pure reagents (abietic acid, palmitic acid, and potassium hydroxide) since they only can grant spectral bands enough resolved to be detected by FTIR spectroscopy together but there is no reason that a similar behavior could not occur in real conditions. Further studies are going on in order to find appropriate technique able to detect this behavior when drying oil and natural resins (colophony, dammar and sandarac) come in play. This would permit to verify if this behavior of terpenic acids take place even in presence of divalent cations (such as zinc or lead) and overall find evidence of resin soaps in metal soaps protrusion in real paintings.

Author Contributions: Conceptualization, T.P.; methodology, T.P., A.P.; validation, E.D., O.C.; formal analysis, T.P., A.P.; investigation, T.P., A.P.; writing—original draft preparation, T.P.; writing—review and editing, T.P.; O.C. and E.D.; supervision, E.D., O.C.; All authors have read and agreed to the published version of the manuscript.

Funding: This research received no external funding.

Institutional Review Board Statement: Not applicable.

Informed Consent Statement: Not applicable.

Data Availability Statement: This study did not report any data.

Acknowledgments: Authors thanks Vanessa Rosciardi and Giuseppina Afruni that with their thesis work helped in the acquiring data process.

Conflicts of Interest: The authors declare no conflict of interest.

References

1. Van den Berg, J.D.J.; Van der Berg, K.J.; Boon, J.J. Chemical changes in curing and ageing oil paints. In Proceedings of the ICOM-CC Triennial 12th Conference, Lyon, France, 29 August–3 September 1999.
2. Boon, J.J.; Peulvé, S.; Van den Brink, O.F.; Duursma, M.C.; Rainford, D. Molecular aspects of mobile and stationary phases in ageing tempera and oil paint films. In *Early Italian Paintings: Techniques and Analysis, Symposium, Maastricht*; Bakkenist, T., Hoppenbrouwers, R., Dubois, H., Eds.; Limburg Conservation Institute: Limburg, The Netherlands, 1997.
3. Van der Weerd, J.; van Loon, A.; Boon, J.J. FTIR studies of the effects of pigments on the aging of oil. *Stud. Conserv.* **2005**, *50*, 3–22. [CrossRef]
4. Boon, J.J.; van der Weerd, J.; Keune, K.; Noble, P.; Wadum, J. Mechanical and chemical changes in old master paintings: Dissolution, metal soap formation and remineralization processes in lead pigmented ground/intermediate paint layers of 17th century painting. In Proceedings of the 13th Triennial Meeting of the ICOM Committee for Conservation, Rio De Janeiro, Brazil, 22–27 September 2002.
5. Noble, P.; Boon, J.J.; Wadum, J. Dissolution, aggregation and protrusion. Lead soap formation in 17th century grounds and paint layers. *Art Matters* **2002**, *1*, 46–61.
6. Robinet, L.; Corbeil, M.-C. The characterization of metal soaps. *Stud. Conserv.* **2003**, *48*, 23–40. [CrossRef]
7. Meilunas, R.J.; Bentsen, J.G.; Steinberg, A. Analysis of aged paint binders by FTIR spectroscopy. *Stud. Conserv.* **1990**, *35*, 33–51.
8. Lazzari, M.; Chiantore, O. Drying and oxidative degradation of linseed oil. *Polym. Degrad. Stab.* **1999**, *65*, 303–313. [CrossRef]
9. Erhardt, D.; Tumosa, C.S.; Mecklenburg, M.F. Natural and accelerated thermal aging of oil paint films. In *Tradition and Innovation: Advances in Conservation*; Roy, A., Smith, P., Eds.; International Institute for Conservation: London, UK, 2000.
10. Van den Berg, J.D.J. *Analytical chemical studies on traditional linseed oil paints*; (MolArt; 6); University of Amsterdam: Amsterdam, The Netherlands, 2002.
11. Colombini, M.P.; Modugno, F.; Fuoco, R.; Tognazzi, A. A GC-MS study on the deterioration of lipidic paint binders. *Microchem. J.* **2002**, *73*, 175–185. [CrossRef]
12. Mallégol, J.; Gardette, J.L.; Lemaire, J. Long-Term Behavior of Oil-Based Varnishes and Paints, I. Spectroscopic Analysis of Curing Drying Oils. *J. Am. Oil Chem. Soc.* **1999**, *76*, 967–976. [CrossRef]
13. Platter, M.J.; De Silva, B.; Gelbrich, T.; Hursthouse, M.B.; Higgitt, C.; Saunders, D.R. The characterization of lead fatty acid soaps in "protrusions" in aged traditional oil paint. *Polyhedron* **2003**, *22*, 3171–3179. [CrossRef]
14. Boon, J.J. Processes inside paintings that affect the picture: Chemical changes at, near and underneath the paint surface. In *Reporting Highlights of the De Mayerne Programme*; Boon, J.J., Ferreira, S.B., Eds.; Netherlands Organisation for Scientific Research (NWO): The Hague, The Netherlands, 2006.
15. Boon, J.J.; Hoogland, F.; Keune, K. Chemical processes in aged oil paints affecting metal soap migration and aggregation. In Proceedings of the AIC annual meeting, Providence, RI, USA, 16–19 June 2006.
16. Keune, K.; Boon, J.J. Analytical Imaging studies of cross-sections of paintings affected by lead soap aggregate formation. *Stud. Conserv.* **2007**, *52*, 161–176. [CrossRef]
17. Keune, K.; van Loon, A.; Boon, J.J. SEM backscattered-electron images of paint cross sections as information source for the presence of the Lead white pigment and lead-related degradation and migration phenomena in oil paintings. In *Microscopy and Microanalysis*; Microscopy Society of America: Reston, VA, USA, 2011.
18. Hermans, J.J.; Keune, K.; Van Loon, A.; Iedema, P.D. Toward a complete molecular model for the formation of metal soaps in oil paints. In *Metal Soaps in Art*; Springer: Cham, Switzerland, 2019.

19. Domenech-Carbo, M.T.; Kuckova, S.; de la Cruz-Canizares, J. Osete-Cortina, Study of the influencing effect of pigments on the photoageing of terpenoid resins used as pictorial media. *J. Chromatogr. A* **2006**, *1121*, 248–258. [CrossRef]
20. Gunn, M.; Chottard, G.; Riviere, E.; Girerd, J.-J.; Chottard, J.-C. Chemical reaction between copper pigments and oleoresinous media. *Stud. Conserv.* **2002**, *47*, 12–23.
21. Poli, T.; Piccirillo, A.; Zoccali, A.; Conti, C.; Nervo, M.; Chiantore, O. The role of Zinc white pigment on the degradation of shellac resin in artworks. *Polym. Degrad. Stab.* **2014**, *102*, 138–144. [CrossRef]
22. Poli, T.; Piccirillo, A.; Nervo, M.; Chiantore, O. Aging of Natural Resins in Presence of Pigments: Metal Soap and Oxalate Formation. In *Metal Soaps in Art*; Springer: Cham, Switzerland, 2019; pp. 141–152.
23. Poli, T.; Piccirillo, A.; Nervo, M.; Chiantore, O. Interactions of natural resins and pigments in works of art. *J. Colloid Interface Sci.* **2017**, *503*, 1–9. [CrossRef]
24. Higgs, S.; Burnstock, A. An investigation into metal ions in varnish coatings. In *Metal Soaps in Art*; Springer: Cham, Switzerland, 2019; pp. 123–140.
25. Carlyle, L. *The artist's assistant: Oil painting instruction manuals and handbooks in Britain 1800-1900, with reference to selected Eighteenth-century sources*; Archetype: London, UK, 2001.
26. Hedley, G. Solubility parameters and varnish removal: A survey. *conservator* **1980**, *4*, 12–18. [CrossRef]
27. Plesters, J. A preliminary note on the incidence of discolouration of Smalt in oil media. *Stud. Conserv.* **1969**, *14*, 62–74.
28. Giovanoli, R.; Muhlethaler, B. Investigation of discoloured Smalt. *Stud. Conserv.* **1970**, *15*, 37–44.
29. Boon, J.J.; Keune, K.; van der Weerd, J.; Geldof, M.; van Asperen de Boer, J.R.J. Imaging microspectroscopic, secondary ion mass spectrometric and electron microscopic studies on discoloured and partially discoloured Smalt in cross-sections of 16 th century paintings. *Chimia* **2001**, *55*, 952–960.
30. Spring, M.; Higgitt, C.; Sauders, D. *Investigation of pigment-medium interaction processes in oil paint containing degraded Smalt*; National Gallery Technical Bulletin 26; Yale University Press: London, UK, 2005.
31. Muhlethaler, B.; Thissen, J. Smalt. *Stud. Conserv.* **1969**, *14*, 47–61.
32. Tournie, A.; Ricciardi, P.; Colomban, P. Glass corrosion mechanisms: A multiscale analysis. *Solid State Ion.* **2008**, *179*, 2142–2154. [CrossRef]
33. Sinclair, R.G.; McKay, A.F.; Jones, R.N. The infrared absorption spectra of saturated fatty acids and esters. *J. Am. Chem. Soc.* **1952**, *74*, 2570–2575. [CrossRef]
34. Jones, R.N.; McKay, A.F.; Sinclair, R.G. Band progressions in the infrared spectra of fatty acids and related compounds. *J. Am. Chem. Soc.* **1952**, *74*, 2575–2578. [CrossRef]
35. Ren, F.; Zheng, Y.F.; Liu, X.M.; Yue, X.Y.; Ma, L.; Li, W.G.; Guan, W.L. An investigation of the oxidation mechanism of abietic acid using two-dimensional infrared correlation spectroscopy. *J. Mol. Struct.* **2015**, *1084*, 236–243. [CrossRef]
36. Nong, W.J.; Chen, X.P.; Liang, J.Z.; Wang, L.L.; Tong, Z.F.; Huang, K.L.; Li, K.X. Isolation and characterization of abietic acid. In *Advanced Materials Research*; Trans Tech Publications Ltd.: Stafa-Zurich, Switzerland, 2014.

Article

Maya Blue Used in Wall Paintings in Mexican Colonial Convents of the XVI Century

Luisa Straulino-Mainou [1,*], Teresa Pi-Puig [2,3,*], Becket Lailson-Tinoco [4], Karla Castro-Chong [4], María Fernanda Urbina-Lemus [5], Pablo Escalante-Gonzalbo [6], Sergey Sedov [2] and Aban Flores-Morán [7]

[1] Coordinación Nacional de Conservación del Patrimonio Cultural, Instituto Nacional de Antropología e Historia (INAH), General Anaya S/N Ex, Av. del Convento, San Diego Churubusco, Coyoacan, Mexico City 04120, Mexico

[2] Instituto de Geología, Universidad Nacional Autónoma de México (UNAM), Cd. Universitaria, Coyoacán, Mexico City 04510, Mexico; sergey@geologia.unam.mx

[3] Laboratorio Nacional de Geoquímica y Mineralogía (LANGEM), Universidad Nacional Autónoma de México (UNAM), Cd. Universitaria, Coyoacán, Mexico City 04510, Mexico

[4] Facultad de Ciencias Sociales y Humanidades, Universidad Autónoma de San Luis Potosí, Alvaro Obregon 64, Centro, San Luis, S.L.P. 78300, Mexico; becket.lailson@uaslp.mx (B.L.-T.); karla.castro@uaslp.mx (K.C.-C.)

[5] Escuela Nacional de Conservación Restauración y Museografía, San Diego Churubusco, Coyoacan, Mexico City 04120, Mexico; fernanda_urbina_l@encrym.edu.mx

[6] Instituto de Investigaciones Estéticas, Cd. Universitaria, Coyoacán, Mexico City 04510, Mexico; pabloeg@unam.mx

[7] Centro de Enseñanza Para Extranjeros, Cd. Universitaria, Coyoacán, Mexico City 04510, Mexico; aflores@cepe.unam.mx

* Correspondence: luisa.straulino@inah.gob.mx (L.S.-M.); tpuig@geologia.unam.mx (T.P.-P.); Tel.: +52-55-5622-4283 (ext. 207) (T.P.-P.)

Citation: Straulino-Mainou, L.; Pi-Puig, T.; Lailson-Tinoco, B.; Castro-Chong, K.; Urbina-Lemus, M.F.; Escalante-Gonzalbo, P.; Sedov, S.; Flores-Morán, A. Maya Blue Used in Wall Paintings in Mexican Colonial Convents of the XVI Century. *Coatings* 2021, 11, 88. https://doi.org/10.3390/coatings11010088

Received: 20 December 2020
Accepted: 12 January 2021
Published: 14 January 2021

Publisher's Note: MDPI stays neutral with regard to jurisdictional claims in published maps and institutional affiliations.

Abstract: Maya blue is a well-known pre-Hispanic pigment, composed of palygorskite or sepiolite and indigo blue, which was used by various Mesoamerican cultures for centuries. There has been limited research about its continued use during the Viceroyalty period; therefore, the sixteenth century is the perfect period through which to study the continuity of pre-Hispanic traditions. The fact that the indigenous people were active participants in the construction and decoration of convents makes their wall paintings a good sampling material. X-ray fluorescence (XRF), scanning electron microscopy (SEM) and X-ray diffraction (XRD) were performed in samples of blue found in convents across Puebla, Tlaxcala and Morelos in order to identify whether the numerous hues of blue were achieved with Maya blue or with other pigments. We found no copper (Cu) or cobalt (Co) with the XRF, so several pigments, such as azurite, smalt or verdigris, were discarded. With SEM, we discovered that the micromorphology of certain blues was clearly needle-shaped, suggesting the presence of palygorskite or sepiolite. In addition, we found silicon (Si), magnesium (Mg) and aluminum (Al) by using energy-dispersive X-ray spectroscopy (EDS) in all blue samples, which also suggests the presence of these magnesium-rich clay minerals. With the XRD samples, we verified that the blues were produced with these two clay minerals, thus confirming that several wall paintings were manufactured with Maya blue. These findings confirm that this particular manmade pre-Hispanic pigment, Maya blue, was an important pigment prior to the Viceroyal period.

Keywords: Maya blue; wall paintings; sixteenth century; palygorskite; X-ray fluorescence; X-ray diffraction

1. Introduction

The arrival of Spaniards in the Mesoamerican territory meant an interaction of two cultures that were unknown to each other. This contact made societies converge in all aspects. Both cultures' presence can be distinguished throughout the XVI century to a greater or lesser extent within institutions, beliefs, buildings and objects. Among these elements, the case of convents stands out since the interaction of cultures took place (more intensely) there, and the creativity in which it was expressed exceeded the limits of what was imagined. The reasons for this convergence were clear: first, in this place, the integration of the indigenous culture into the western was carried out in a planned way; second, the indigenous people, being the ones who constructed these buildings, portrayed their cosmovision in them. The clearest example of this is in the wall painting decoration since jaguars, eagles, native plants and pre-Hispanic symbols interact with the Christian saints and scenes of Jesus' life. This interaction also occurred in the techniques used for this type of decoration, since the indigenous people who were painting the walls introduced their techniques and materials. Thus, we can see in the murals how the oxide red, the black smoke, the lepidocrocite and the Mayan blue produced an Indigenous–Christian image [1–6] (Figure 1).

Figure 1. Examples of different hues of blues used in XVI century wall painting. From these wall paintings, very small samples were taken for XRF, SEM-EDS and XRD: (**a**) Tezontepec, (**b**) Ocuituco, (**c**) Oaxtepec, lower cloister and (**d**) Oaxtepec, upper cloister.

Among these colors, Maya blue is a special case. Maya blue is a synthetic pigment made from a fibrous clay mineral (palygorskite, with international mineralogical association (IMA) approved formula $(Mg,Al)_2Si_4O_{10}(OH)\cdot4H_2O$ [7]), plus an organic colorant (indigo: $C_{16}H_{10}N_2O_2$) [8]. There is crystallographic evidence that the Maya blue from Central Mexico used a different palygorskite compared to the Yucatan one. This is because Yucatecan palygorskite is a mixture of almost equal parts of monoclinic and orthorhombic palygorskite, while the Maya blue of Central Mexico exhibits mostly the orthorhombic form. However, some studies conducted in Central Mexico have identified a different Maya blue pigment made with sepiolite instead of palygorskite in archaeological objects [9].

Maya blue was used from the fifth century AD to colonial times in Mexico; nevertheless, there have been some studies that placed Maya blue in Cuba during its colonial period, identified as the pigment in a color formerly known as "Havana blue". This blue pigment was used mostly in decorations dated from the middle of the eighteenth century to around 1860 [10,11].

There has been some evidence of the use of Maya blue in XVI century codices such as *La Historia General de las Cosas de la Nueva España* by Fray Bernardino de Sahagún [12], in some maps (Ameca, Atlatlauca, Ixtapalapa, Mextitlán and Tehuantepec) of the *Relaciones Geográficas* made between 1578 and 1583 [13] and in the Cuauhtinchan 2 Codex, in which the indigo colorant was detected with Raman spectroscopy and the palygoskite with Fourier transform infrared spectroscopy (FTIR) [14], as some examples. Additionally, some cases of Maya blue used in XVI century convents have been identified in Jiutepec (Morelos), Actopan (Hidalgo), Epazoyucan (Hidalgo), Cuahutinchan (Puebla), Tezontepec (Hidalgo) and Totimehuacán (Puebla) [8,9,15].

Furthermore, some non-invasive studies, such as those using UV lighting, surface-enhanced Raman spectroscopy (SERS), X-ray fluorescence and FTIR, have been conducted on XVI century convents of the Augustinian order. With these techniques, indigo was found in the wall paintings of Actopan, Epazoyucan and Ixmiquilpan; nevertheless, Wong (2014) reported that the data acquired with FTIR were not of sufficient quality and thus they were not used; in consequence, it then was not possible to identify palygorskite, and therefore Maya blue could be the pigment used for the blues that exhibited indigo [16,17].

The search for Maya blue is further complicated by the fact that a number of other blue pigments were used in colonial times. Some of these pigments were introduced by the Spanish people and others were known in pre-Hispanic times. Among them are ultramarine (lazurite mineral), azurite (copper carbonate) and smalt blue (glass with cobalt). Some other blues were incorporated later on with the development of chemistry in the XVIII and XIX centuries, such as Prussian blue (ferrous ferrocyanide), cobalt blue (cobalt aluminate), cerulean blue (cobalt stannate), Bremen blue (copper hydroxide with copper carbonate) and manganese blue (manganate crystals mixed with barium sulfate) in the XX century [18–20]. These pigments could not only substitute but also occur together with the Maya blue due to repainting or restorations.

The aim of this paper is to identify the Maya blue pigment in the blue hues on several XVI century wall paintings in Central Mexico and to discriminate it from other pigments with similar tonalities. We propose to use palygorskite and/or sepiolite as an indicator of Maya blue's presence, whereas high concentrations of some specific elements (Cu, Co, Fe, etc.) could evidence the use of different pigments. To obtain these indicators, we applied the following methods: portable X-ray fluorescence equipment, scanning electron microscopy with energy-dispersive spectroscopy and X-ray diffraction. All of these techniques are well established to characterize pigments used in wall paintings [21–24]. The presence of these clay minerals in blues, in combination with other characteristics of the samples (such as absence Cu or Co minerals), is an almost certain indication of the used pigment being Maya blue. Moreover, we acquired information about the pictorial technique in which the Maya blue was applied.

2. Materials and Methods

The study was focused on the Central Highlands of the current Mexican territory, on the convents founded by the mendicant orders in the states of Hidalgo, Tlaxcala, Morelos, Puebla, Estado de México and Mexico City. This territorial delimitation is due to the fact that, particularly in these locations, the mendicant orders (especially the Franciscans and Augustinians) sought to integrate indigenous cultures into the new developing society. On the other hand, the region was inhabited by groups affiliated with the Nahua and Otomí, which had a particular way of creating a unique image and cosmovision, thus producing art that was different from that seen in Oaxaca, Michoacán or the Mayan area. With this delimitation, we restricted the study to 67 foundations which had wall painting remains—from small fragments to complete programs (Appendix A shows all convents with wall paintings in the Central Highlands of Mexico). Due to time constraints, only 35 were broadly studied with XRF, and for conservation purposes, small samples were taken only in 10 of them.

2.1. X-ray Fluorescence (XRF)

The palette colors of 35 convents were studied with the portable XRF equipment, collecting preliminary data before taking physical samples to define the best locations for sampling. In these 35 convents, we observed a palette of 13 colors with different hues plus the color given by the plasters and stuccos. Twenty-eight of the convents had blue colors, and 2 had greenish blues, possibly composed of Maya blue (Table 1).

Table 1. List of convents whose paintings were studied by XRF, their location, foundation year and color of pigments sampled.

Convent	State	Year	st	ye	or	bl	br	cr	fl	gr	pr	bl	rd	bg	pk	gr	pl
1. Yecapixtla	Morelos	1535	x	x	–	x	x	–	–	–	–	x	x	x	–	–x	x
2. Atlatlahucan	Morelos	1569	x	–	x	–	–	–	–	–	–	x	–	x	–	x *	x
3. Izúcar de Matamoros	Puebla	1540	x	–	x	–	x	–	x	–	–	x	x	x	–	x	–
4. Atlixco	Puebla	1550 (c.)	x	–	x	x	–	–	–	–	–	x	x	–	–	–	–
5. Ixmiquilpan	Hidalgo	1550	x	x	x	x	–	–	x	–	–	x	x	–	–	x	x
6. Tochimilco	Puebla	1569	–	–	x	–	–	–	x	–	–	x	x	–	–	x	–
7. Cuernavaca	Morelos	1552	x	–	x	x	x	–	–	–	–	x	x	–	–	x	–
8. Oaxtepec	Morelos	1534	x	x	–	x	x	–	x	–	x	x	x	–	–	x	x
9. Tepoztlán	Morelos	1559	x	x	–	x	x	–	–	–	–	x	x	–	–	x	x
10. Ocuituco	Morelos	1536	x	–	x	x	–	–	x	x	–	x	–	x	x	–	–
11. Zacualpan de Amilpas	Morelos	1535	x	–	x	x	–	–	–	–	–	x	–	x	–	–	x
12. Tetela del volcán	Morelos	1561	x	x	x	–	x	–	–	–	–	x	–	x	–	x	x
13. Tepeji del Río	Hidalgo	1558	x	–	–	x	x	–	–	–	–	x	x	–	x	x	–
14. Alfajayucan	Hidalgo	1559	–	x	x	x	–	x	–	–	–	x	x	–	–	x	x
15. Actopan open chapel	Hidalgo	1550	x	x	x	x	–	–	–	–	–	x	x	–	–	–	x
16. Tezontepec	Hidalgo	1554	x	x	x	x	x	–	–	–	–	x	x	–	–	–	x
17. Tepetitlán	Hidalgo	1571	–	x	–	x	x	–	x	–	x	–	–	–	–	x	–
18. Actopan convent	Hidalgo	1550	x	x	x	x	–	–	–	–	–	–	x	–	–	–	–
19. Atotonilco el grande	Hidalgo	1536	x	x	–	–	–	–	–	–	–	x	x	–	–	–	x
20. Acatlán	Hidalgo	1569	x	x	–	x	–	–	–	–	–	x	x	x	–	–	–
21. Yautepec	Morelos	1550	x	–	–	–	–	–	–	–	–	x	x	–	–	–	x
22. Tlaltizapán	Morelos	1576	x	–	x	x–	x	–	x	–	–	x	x	x	–	–	x
23. Tlaquiltenango	Morelos	1590	x	x	x	x	x	–	–	–	–	x	x	–	–	–	x
24. Calpulalpan	Tlaxcala	1576	x	x	–	–	x	–	–	–	–	x	–	–	–	–	–
25. Tepeapulco	Hidalgo	1529	x	x	x	x	–	–	–	–	–	x	–	x	–	x	x
26. Zempoala	Hidalgo	1569	x	x	–	x	x	–	–	–	–	x	x	–	–	–	x
27. Epazoyucan	Hidalgo	1540	x	–	x	x	x	–	–	–	–	x	x	–	–	–	x
28. Oxtoticpac	Estado de México	1570	x	–	x	x	x	–	–	–	–	x	x	–	–	–	x
29. Cholula	Puebla	1552	x	x	x	x	x	–	x	–	–	–	x	x	–	x	–
30. Huejotzingo	Puebla	1530	–	x	x	x	x	–	–	–	–	x	–	x	x	x	–
31. Huatlatlauca	Puebla	1566	x	x	x	x	x	–	x	–	–	x	x	–	–	x	x
32. Tecalli	Puebla	1554	x	x	–	x	x	–	x	–	–	x	–	x	–	x	
33. Cuahutinchan	Puebla	1554	x	x	x	x *	x	–	–	–	x	x	x	–	–	x *	x
34. Tula	Hidalgo	1543	x	x	–	x	x	–	–	–	–	x	x	–	–	–	–
35. Meztitlán	Hidalgo	1537	x	x	–	x	–	–	–	–	–	x	x	–	–	x *	–

st (stucco), ye (yellow), or (orange), bl (blue), bl * (greenish blue), br (brown), cr (cream), fl (flesh, gr (gray), pr (purple), bl (black), rd (red), bg (burgundy), pk (pink), gr (green), gr * (bluish green), pl (plaster). The colors that can be related to Maya blue are marked in bold and highlighted in gray.

X-ray fluorescence spectroscopy is a technique used to determine the elemental composition of materials. The technique is designed to detect the interactions of radiation with matter, i.e., the photoelectric effect of X-rays, which are characteristic of the elements that constitute a material [25]. The sample under analysis is first excited by a primary, high-energy X-ray; this causes the emission of characteristic X-rays of lower energy (fluorescence), which serve as a spectroscopic fingerprint for each element present in the sample.

By counting the number of photons within a certain emitted energy, the elements that compose the sample can be identified and quantified [26].

Portable XRF analysis is relevant to field research, as it allows for preliminary and non-invasive identification of materials, aiding in proper sample selection and reducing aggressive sampling. The portable XRF equipment has a compact system with a working voltage of 40 to 60 kV, with variable current in the micro amp range. It has been demonstrated that portable XRF analysis is an accurate and fast technique for the compositional characterization of a sample [27]. However, this type of equipment only allows for superficial examination of the sample, since the emitted energy is not high enough to penetrate subjacent layers. The scope of this type of equipment in the characterization of pictorial materials is limited, due to the fact that they often involve different layers and thicknesses. Therefore, the analysis of pictorial materials should be supplemented with other archaeometric techniques.

The XRF analysis was performed with an Innov-X Omega OSD2000 X-Ray handheld analyzer (Innov-X Systems, Inc., City, MA, USA), provided and available- by the Archaeology Laboratory of the Faculty of Social Sciences and Humanities at Universidad Autonoma de San Luis Potosí. The excitation source consists of an X-ray tube with large area Silicon Drift Detector (SDD). The SDD provides 10× improvement in signal to background ratio, marked resolution improvement, and it operates at 195 to 165 eV (it has the capacity to handle 10× more counts).

In this study, the analyzer window was placed directly in specific sections of the selected mural paintings in situ (according to the colors of interest) and each sample was irradiated for 2 min. The device was configured in MINING Application (PiN Detector: XPD6000, Innov-X Systems, Inc., City MA, USA) with a voltage of 40 kV and a current of 20 mA. This equipment detects up to 30 elements, with accuracy greater than 90% for calcium (Ca), iron (Fe), silicon (Si), magnesium (Mg) and aluminum (Al).

Before performing tests, it was necessary to standardize the instrument; thus, the analyzer hardware was initiated or restarted every time the instrument was operated for more than four hours. The procedure consists of placing the 316 stainless steel standardization clip on the analyzer nose and tapping the standardization (STD) function. It takes 20 s for the analyzer to calibrate to the parameters of the mode that has been selected for the tests. The automated standardization allows for the collection of a spectrum on a known standard (Alloy 316); it also involves the comparison of a variety of parameters to values stored when the instrument was calibrated at the factory.

The Innov-X Omega analyzer was set in MINING mode, in order to detect only metallic elements that might be present in the pigments. In addition, this analytical option helped to reduce the spectrum of the elements detected in the analysis and to characterize the typology of pigments, since the color hue of each pigment is given by a specific metal (iron, copper, manganese, etc.). Thus, this method allowed us to conduct an extensive preliminary in situ investigation of color and pigments, which then served to aid in our selection of locations and colors for physical sampling to further analyze them with SEM-EDS and XRD [28–30].

2.2. Scanning Electron Microscopy–Energy Dispersive X-ray Spectroscopy (SEM-EDS)

We selected 12 blue samples of 10 convents (Table 2) to be analyzed with a JEOL JSM6060LV (JEOL USA, Inc., Peabody, MA, USA) equipped with an INCA Energy 250 EDSLK-IE250 (Oxford Instruments, Oxfordshire, UK). The images were obtained on backscattering (BSE) mode in the surface and in the cross-sections of micro-fragments, under low vacuum, with 20 kV and diverse magnifications. Elemental composition was acquired using EDS in various modes: punctual, area and map distribution of chemical elements. The criteria described in Piovesan et al. [22] were followed to determine the painting technique.

Table 2. Number of blue samples that were taken from each convent, the location of the convent, their foundation year and their monastic affiliation.

Convent	State	Year	Monastic Order	Number of Samples	Hue
Atlatlahucan	Morelos	1569	Augustinian	1	Greenish blue
Cuahutinchan	Puebla	1554	Franciscan	2	Light blue/dark blue
Cuernavaca	Morelos	1552	Franciscan	1	Blue
Ixmiquilpan	Hidalgo	1550	Augustinian	1	Blue
Meztitlán	Hidalgo	1537	Augustinian	1	Greenish blue
Oaxtepec	Morelos	1534	Dominican	2	Blue
Ocuituco	Morelos	1536	Augustinian	1	Blue
Tepeapulco	Hidalgo	1529	Franciscan	1	Blue
Tlaltizapán	Morelos	1576	Dominican	1	Blue
Tezontepec	Hidalgo	1554	Augustinian	2	Light blue/dark blue

2.3. X-ray Diffraction (XRD)

Five samples previously analyzed with SEM-EDS were selected for X-ray diffraction (Table 3). These samples were chosen according to two different criteria: needle-shaped minerals were not clearly detected with SEM in the pigment layer; thus, it was important to ensure a clear identification of the mineral component of the blue pigment and/or the blue compositions showed relatively high quantities of Fe.

Table 3. Blue samples analyzed by XRD, showing a microscopic image and the location of the sample in each convent.

Sample	Image with Optical Microscope	Photograph with the Blue Sample Location
Blue/Ocuituco (M1)		
Light Blue/Tezontepec (M2)		
Dark Blue/Tezontepec (M5)		
Blur/Oaxtepec upper cloister (M3)		
Blue/Oaxtepec lower cloister (M4)		

The samples sizes are very small and mainly consist of detached fragments due to conservation and displaying concerns. These wall paintings are taken from museums or convents still in use; thus, sampling should be restricted.

The samples were homogenized using an agate mortar and mounted on a zero-background sample holder due to their very small size.

X-ray diffraction spectra of non-oriented aliquots have been acquired using an Empyrean diffractometer (Malvern Panalytical, Malvern, UK), operating with an accelerating voltage of 45 kV and a filament current of 40 mA, and using CoKα radiation, nickel filter and a PIXcel 3D detector (Malvern Panalytical, Malvern, UK). Samples were measured in the range of 4°–80° (2θ) with a step size of 0.002° (2θ) and 90 s of scan step time. Phase identification and quantification by the Rietveld method [31] were completed using the Highscore v4.5 software (Malvern Panalytical, Malvern, UK), as well as ICCD (International Center for Diffraction Data) and ICSD (Inorganic Crystal structure Database) databases.

In a second stage, the samples were heated to 450 °C and measured by X-ray diffraction under the same conditions as the original samples, for the purpose of verifying the destruction of the basal peaks of palygorskite (~10.3 Å) and sepiolite (~12 Å) and, in turn, confirming the presence of these mineral phases [32].

3. Results

The next paragraphs show the results obtained with each technique. Overall, results showed that Maya blue could have been widespread in XVI century wall painting in convents.

3.1. XRF. Concentration of Selected Elements in Blue Pigments

Metal elemental compositions of blues were compared with the metal elemental composition of stucco where possible (see Appendix B, where all the samples and analyzed elements are shown); Huejotzingo, Tepetitlán and Alfajayucan had a white painting coating over almost every space of stucco, preventing the acquisition of high-quality information.

Some blues used in colonial times (see Introduction) and other pigments added in the XIX and XX centuries could be found in the convents' wall paintings, all of them having a distinct composition. Thus, XRF analysis allowed us to perform a quick study of the composition of blues and, consequently, to devise a hypothesis of the most likely used pigment.

Regarding the amount of iron (Fe) and the possibility of Prussian Blue repaints, results show that 12 hues of blues have a greater increase in Fe compared with the quantity in the stuccos, 14 hues of blues have a slight increase in Fe, while 5 have almost the same amount and 3 have less Fe in comparison to the corresponding stucco.

Copper (Cu) was found only in three samples (the samples from Tepeji del Río and one sample from Cholula) in ranges between 0.01% and 0.034% while the corresponding stucco had no Cu (Figure 2). It is possible that the blues found in Tepeji are made of a copper-based pigment while the sample from Cholula (0.01% of Cu) could be a mixture of blue pigments including a copper-based one.

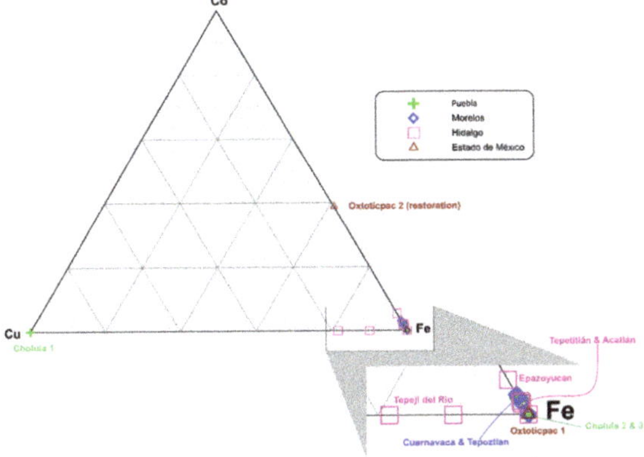

Figure 2. Fe–Cu–Co triangular diagram of the studied blue pigments from convents of different states in Mexico. Data acquired with EDS.

Regarding cobalt (Co), two blue samples (Atlixco and Zacualpan de Amilpas) have Co in their compositions. However, the stucco had similar quantities of this element; hence, it is not probable that the blue sample is a cobalt-based blue. There were, nonetheless, other samples that exhibited Co while their corresponding stuccos did not. These samples were one sample from Cuernavaca, the sample from Tepoztlan (which also exhibited a high quantity of Fe), one sample from Tepetitlán, the sample from Acatlan, one sample from Epazoyucan and the sample from Oxtoticpac, clearly recognized as a restoration. These pigments could be cobalt-based pigments, and as such, they could correspond to later (XVIII or XIX century) repainting or to restoration interventions. The sample of Tepoztlán could be a mixture of Prussian blue and a cobalt-based pigment.

The sample from Oxtoticpac mentioned above (a clear restoration intervention), as well as the analyzed areas of Tepeji del Río, showed clear differences between the global composition when compared with the other blues.

3.2. SEM-EDS. Painting Layer Micromorphology

It can be noticed in all the scanning electron microscopy images (Figures 3 and 4) that the color is clearly in a different layer than the stucco and plaster; therefore, none of these blues were applied using a fresco technique. All were painted with a dry technique using an organic binder (see Table 4), except for the Tezontepec and Cuernavaca samples, where the pigment appears to be mixed with gypsum and lime in a technique similar to lime painting but also incorporating calcium sulfate.

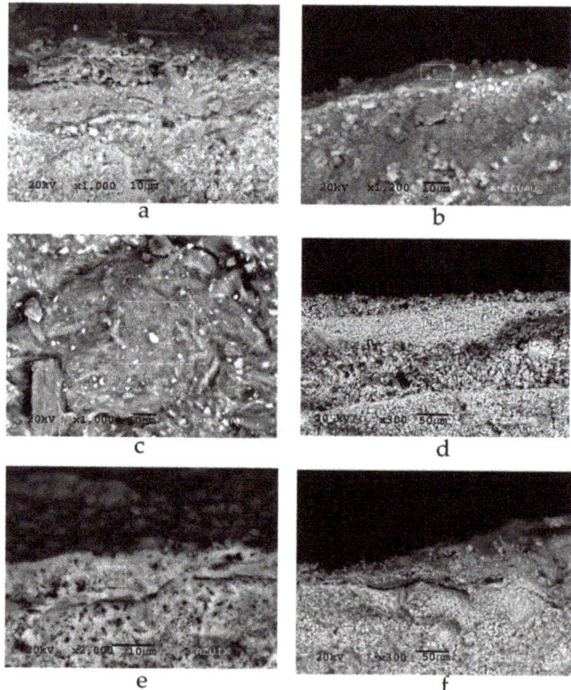

Figure 3. SEM images of different blue samples. (**a**) Atlatlaucan cross-section with three main layers. The one at the top is the paint layer that is placed over a stucco and, on the bottom, the plaster can be seen. (**b**) Cuauhtinchan, cross-section with three main layers. The one at the top is the paint layer that is placed over a thin white layer of stucco placed over a plaster. (**c**) Cuauhtinchan, surface of the pain layer, a mesh of thin needle-shaped features can be seen. (**d**) Cuernavaca cross-section. The pigment is dispersed within the surface layer, where dark areas can be observed. The pictorial layer is placed over several layers of gypsum. (**e**) Ixmiquilpan, cross-section. The top layer is the blue paint placed over a thin white layer made of gypsum, and on the bottom, the plaster is observed. (**f**) Meztitlán, cross-section. The paint layer is placed directly over a plaster.

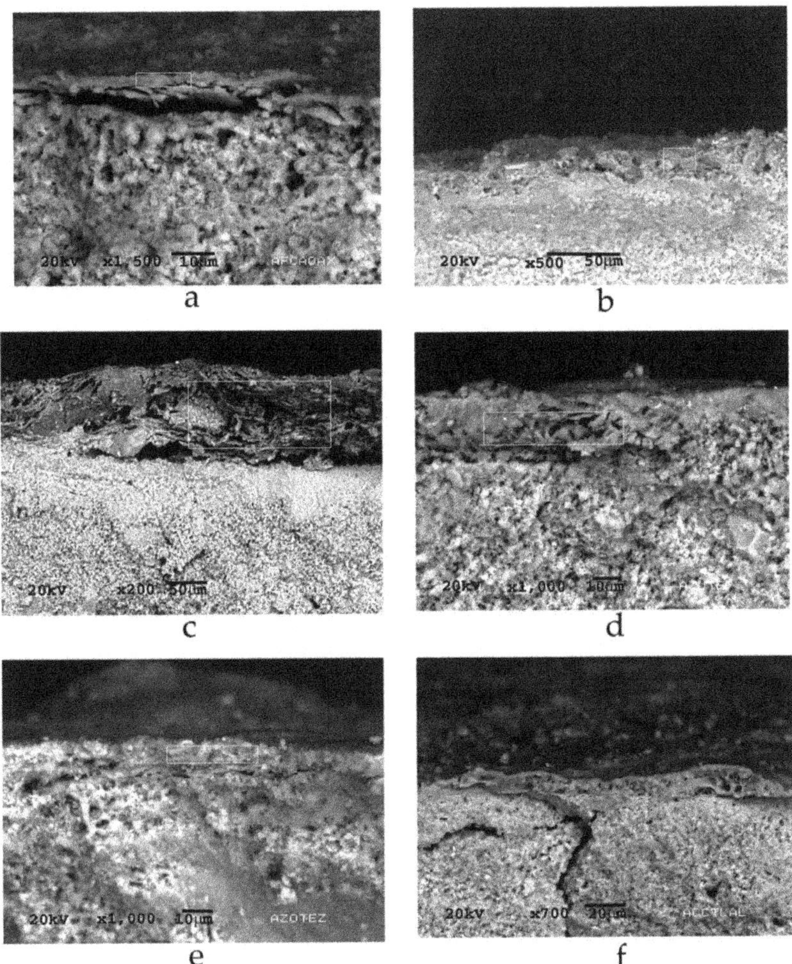

Figure 4. SEM images of different blue samples. (**a**) Oaxtepec upper cloister, cross-section with three main layers. The one at the top is the paint layer that is placed over a stucco barely distinguishable from the plaster by its coarser and more porous texture. (**b**) Oaxtepec lower cloister, cross-section. The dark layer on top is the blue paint placed over a stucco, and on the bottom, a less porous plaster is observed. (**c**) Ocuituco, cross-section. The blue paint layer is thick and placed over a compact stucco layer; at the bottom, a more porous plaster is observed. (**d**) Tepeapulco, cross-section. A thin dark layer of blue paint is placed over a stucco; then, the more compact plaster is seen. (**e**) Tezontepec, cross-section. A thin dark layer of blue paint is placed over a thin stucco that is clearly separated from the plaster on the bottom. (**f**) Tlaltizapan, cross-section. The paint layer is directly placed over the plaster.

Table 4. Elemental analysis of blue samples with EDS. The elements and percentage are shown in the columns. Ca and S concentrations are related to the presence of gypsum that was identified with XRD. The high concentration of C is related to the use of an organic binder in the paint.

Convent	Sample	C	O	Na	Mg	Al	Si	P	S	Cl	K	Ca	Fe	Cu	Pb	Total
Atlatlauhcan	Greenish blue	**22.24**	48.24	–	0.62	1.35	5.03	–	**5**	**0.32**	**0.32**	16.14	0.74	–	–	100
Cuauhtinchan	Blue	**19.59**	49.54	–	1.55	2.24	6.15	–	**6.13**	–	–	14.79	–	–	–	100
Cuauhtinchan	Blue lateral wall	**12.61**	51.62	–	8.88	0.42	19.43	–	**0.77**	**0.15**	–	4.71	1.09	0.31	–	100
Cuernavaca	Blue	**42.61**	41.98	–	-	0.23	0.74	–	**2.71**	**11.73**	–	–	–	–	–	100
Ixmiquilpan	Blue	**16.04**	45.17	–	0.25	0.16	1.39	–	**4.2**	**0.25**	–	31.68	–	–	0.85	100
Meztitlán	Blue	**24.07**	46.78	–	1.34	1.34	9.37	**2.15**	**1.65**	**0.13**	**0.29**	12.57	0.31	–	–	100
Oaxtepec	Blue upper cloister	**29.5**	50.53	**0.14**	0.41	0.63	3.31	**0.49**	**3.31**	**0.36**	**0.31**	10.87	0.14	–	–	100
Oaxtepec	Blue lower cloister	**27.12**	46.42	–	1.11	0.89	5.567	**0.42**	**1.15**	–	**0.59**	16.09	0.54	–	–	100
Ocuituco	Blue	**18.5**	55.89	–	1.96	0.35	4.8	–	**5.3**	–	**0.3**	12.9	–	–	–	100
Tepeapulco	Blue	**22.51**	50.96	–	2.79	1.19	8.24	–	**1.57**	**0.34**	**0.24**	11.73	0.42	–	–	100
Tezontepec	Dark blue	**45.54**	42.53	**0.34**	0.19	0.51	1.17	–	**3.4**	**0.19**	**0.06**	5.89	0.18	–	–	100
Tezontepec	Light blue	**52.27**	35.93	**0.21**	0.2	0.15	0.63	–	**3.77**	–	–	6.64	0.2	–	–	100
Tlaltizapan	Blue	**25.47**	49.62	**0.51**	0.96	0.8	9.93	**0.36**	**0.89**	**0.26**	**0.51**	10.44	0.25	–	–	100
–	Egg (mixed white and yolk)	**85.67**	12.49	**0.16**	–	–	–	**0.54**	**0.57**	**0.36**	**0.21**	–	–	–	–	100

Elements related to palygorskite and sepiolite are highlighted with grey shading, while elements possibly related to an organic binder are marked in bold.

All samples have a clear stucco or preparation base (a thin layer between the plaster and the colour layer) below the painting layer, with the exception of the samples from Meztitlán and Tlaltizapan. In the first one, the blue color is placed directly on a lime plaster, while in the latter, a gypsum–lime plaster is the base. In the other samples, the blue colors are placed on a gypsum–lime stucco or preparation base or on stucco made only from gypsum; this is clearly a technique introduced by Spaniards considering that, to date, there is almost no scientific information that indicates the use of gypsum in wall painting in pre-Hispanic times. In Figure 3c, this corresponds to the blue lateral wall sample from Cuauhtinchan, in which a net of a mineral with a needle-shaped morphology corresponding to the palygorskite can clearly be seen.

On the other hand, to estimate the elemental composition of the pictorial layer, the samples were analyzed in cross-sections with EDS and focused area analyses were carried out only at this layer; the complete results are shown in Table 4. In all studied areas of all the samples, silica (Si), aluminum (Al) and magnesium (Mg) elements were found, and these elements correspond to the composition of palygorskite, except for the sample from Cuernavaca, in which no traces of Mg or Fe were found. Several of them also present iron in their composition.

On the SEM images of the samples, some physical deterioration can be observed, including fissures, fractures and detachments of the paintings. The sampling method, in which some pressure had to be applied to separate the minimal samples, could have caused some of these features. Nonetheless, major chemical changes are not expected in the mineral fractions of the paint layer, the stucco or plasters of the studied samples because no morphological evidence of dissolution or alteration of original substances was found, such as etch pits and synthesis of neoformed components as secondary carbonates. Usually, such components have quite specific shapes, showing a sharp difference from the original constituents of stucco and pigment layers. We also did not find any biological agents of deterioration—cells of cianobacteria, fungi, etc. This is partly due to the fact that for the SEM/EDX analysis, we selected fragments with minimal macroscopic deterioration features. Thus, supposedly, the chemical signal from mineral components from EDS microanalysis originates predominantly from the primary components and is not influenced significantly by posterior alteration.

To identify unequivocally these minerals, it would be ideal to calculate their structural formula from chemical point analysis obtained with a scanning electron microprobe. However, given the porous nature of the material and the small amount of pigment available, this type of analysis was not possible, and we had to use SEM-EDS data to be able to chemically classify them without calculating their structural formula.

It is important to mention that published results of microanalyses of individual particles of both sepiolite and palygorskite are quite rare; this is because these results can be affected by other clay minerals and other associated minerals, such as impurities [33]. Many of the palygorskite samples contain impurities and, thus, obtaining an accurate analysis is always difficult. This is why it was decided to use the triangular diagram (Figure 5) of the AFM type (Al_2O_3–Fe_2O_3–MgO) to compare the analyzed samples (SEM-EDS) of blue pigments from the different convents with each other—and mainly with the palygorskite and sepiolite reported in the mineralogical literature [33]. Some palygorskite samples reported in Maya blue-type pigments from the Mayan zone were also included [34].

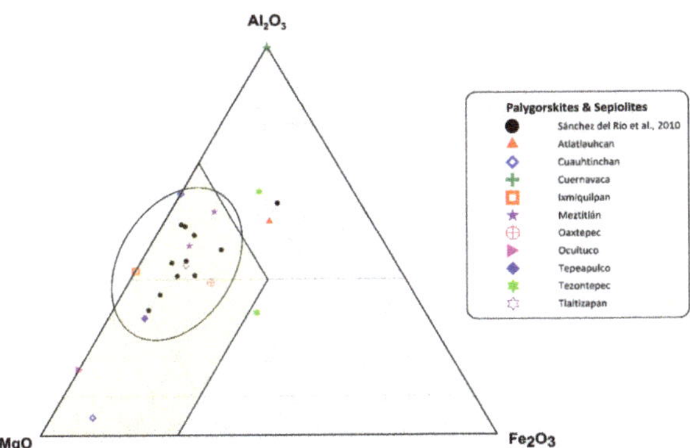

Figure 5. Triangular AFM diagram. In yellow, area determined for palygorskite and sepiolite samples studied by García-Romero & Suárez [33].

With the exception of the samples from Atlatlahucan (slightly enriched in aluminum), Tezontepec (slightly enriched in aluminum and iron) and Cuernavaca (without iron and magnesium), all other samples of blue pigment are projected within the field established for the sepiolite–palygorskite group [33]. Additionally, most of them—with the exception of a sample from Cuauhtinchan and Ocuituco—are very close to the samples reported by Sánchez del Río et al. [34], for Maya blue-type pigments collected in the area of the Yucatán Peninsula. These last two pigment samples could correspond to sepiolite, since this mineral occupies the most magnesic and trioctahedral extreme, while palygorskite occupies the most aluminic–magnesic and dioctahedral extreme and has a lot of vacancies. It is important to mention that, unlike what many authors have indicated in the past, these authors [33] show that there is no compositional gap between both mineralogical species.

Moreover, the phosphorus, potassium, chlorine and sodium found with EDS can correspond to some elements present in the binders mixed with the pigments. Collagen and egg, which were commonly used to paint in secco techniques, usually present these elements. In Figure 6, we compare the elemental composition of egg (mixture of whites and yolk) and some blue samples.

As it can be observed in the graph, the elemental composition matches well with a proteic binder, such as egg; nevertheless, further analyses should be carried out to determine the exact nature of the binder.

Sulfur is high in the blue samples and this aspect can be related to the preparation of stuccos or preparation bases, which contain or are completely made of gypsum. In addition, gypsum was added as a binder in the painting layer in some cases. Phosphorus is also high if compared with the quantity present in the egg in the Meztitlán sample. This sample has a thin, black layer in some areas over the blue. Thus, this black could be made from ash or carbonized bone.

The samples from Cuernavaca also exhibit a high quantity of chlorine, which can possibly be correlated to the presence of salts within the wall paintings. In one of these samples from Cuernavaca, a sodium nitrate salt was found (Figure 7), indicating a conservation problem caused by saline efflorescence.

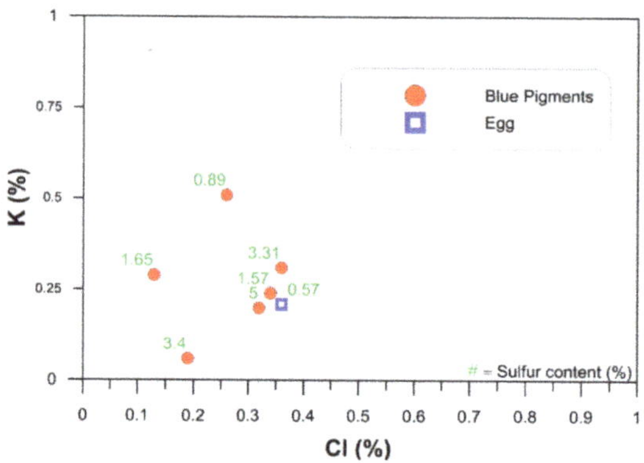

Figure 6. Comparison between elemental composition (K, Cl, S) of blue samples and hen egg (yolk and white, mixed).

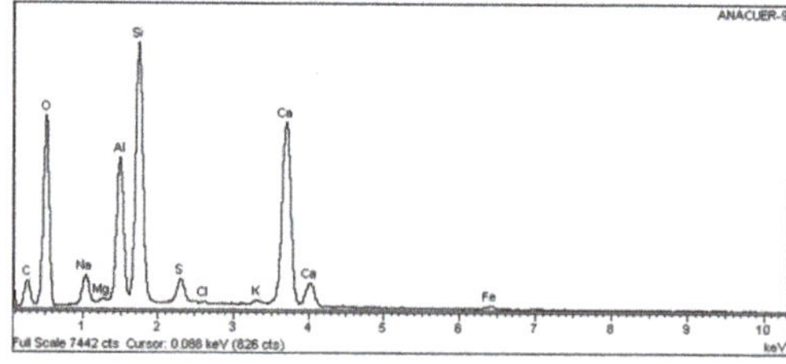

Figure 7. Efflorescence and their elemental composition found in a sample of Cuernavaca.

3.3. XRD. Mineralogical Composition of the Pigments

Five samples (M1 to M5) were characterized. By means of X-ray diffraction analysis, the presence of sepiolite in one sample (M1), and of palygorskite in the other four samples (M2, M3, M4 and M5), could be confirmed. The mineralogical results are shown in Table 5. The identification of the sepiolite was based on the presence of peaks with interplanar distance of 11.9 and 4.45 Å. The identification of palygorskite was carried out based on the presence of peaks with interplanar distance of 10.36 and 4.46 Å (Figure 8).

Table 5. XRD results. Quantitative results by Rietveld method could only be obtained for samples M1 and M5. GOF = goodness of fit. IMA (International Mineralogical Association) approved formulas of minerals (http://cnmnc.main. jp/) are indicated.

Sample	Identified Phases	XRD Patterns	Quantitative Rietveld (%)	GOF
M1 Blue/Ocuituco	Calcite: $CaCO_3$ Gypsum: $CaSO_4 \cdot 2H_2O$ Bassanite: $CaSO_4 \cdot 0.5H_2O$ Sepiolite: $Mg_4Si_6O_{15}(OH)_2 \cdot 6H_2O$	ICDD 01 072 4582 ICSD 98 016 1623 ICDD 01 074 2787 ICDD 98 003 1142	32.6 (7) 36.9 (5) 8.6 (9) 21.9 (9)	0.967
M2 Light Blue/ Tezontepec	Calcite: $CaCO_3$ Gypsum: $CaSO_4 \cdot 2H_2O$ Palygorskite: $(Mg,Al)_2Si_4O_{10}(OH) \cdot 4H_2O$	ICDD 01 072 4582 ICSD 98 016 1623 ICDD 98 004 0688	–	–
M3 Blue/Oaxtepec upper cloister	Calcite: $CaCO_3$ Gypsum: $CaSO_4 \cdot 2H_2O$ Palygorskite: $(Mg,Al)_2Si_4O_{10}(OH) \cdot 4H_2O$	ICDD 01 072 4582 ICSD 98 016 1623 ICDD 98 004 0688	–	–
M4 Blue/Oaxtepec lower cloister	Calcite: $CaCO_3$ Palygorskite: $(Mg,Al)_2Si_4O_{10}(OH) \cdot 4H_2O$	ICDD 01 072 4582 ICDD 98 004 0688	–	–
M5 Dark Blue/ Tezontepec	Calcite: $CaCO_3$ Gypsum: $CaSO_4 \cdot 2H_2O$ Bassanite: $CaSO_4 \cdot 0.5H_2O$ Palygorskite: $(Mg,Al)_2Si_4O_{10}(OH) \cdot 4H_2O$	ICDD 01 072 4582 ICSD 98 016 1623 ICDD 01 074 2787 ICDD 98 004 0688	43.2 (8) 33.1 (9) 10.4 (7) 13.2 (8)	0.953

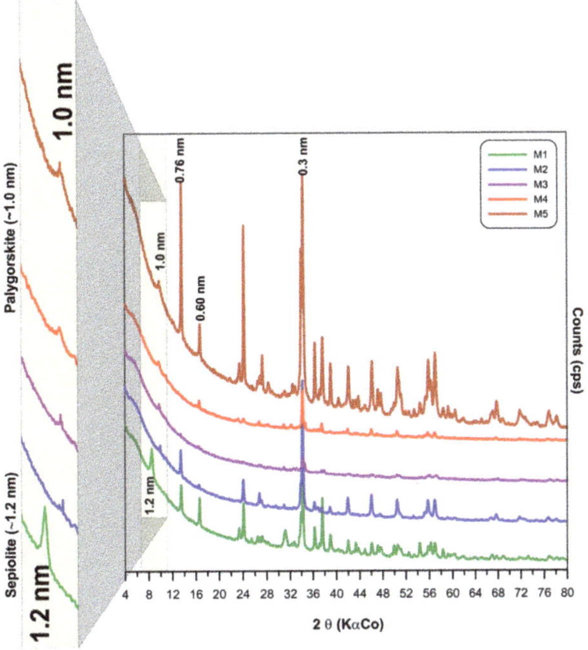

Figure 8. XRD patterns of the five measured samples (M1, M2, M3, M4 and M5). The samples were measured using a fine focus cobalt tube (KαCo radiation). The basal peaks with interplanar distance of 1.0 and 1.2 nm enlarged at the left of the figure are characteristic of palygorskite and sepiolite, respectively.

Paligorskite and sepiolite are minority phases in all samples due to the presence of abundant gypsum, bassanite and non-magnesian calcite in the paste, showing that the blue pigmented layer is very thin (Table 5). Quantification by the Rietveld method (Table 5) could only be applied to samples M1 and M5 because the other samples were very small, and the signal-to-noise ratio does not allow us to obtain good fit indices. Figure 9 shows the result of the Rietveld refinement for samples 1 and 5.

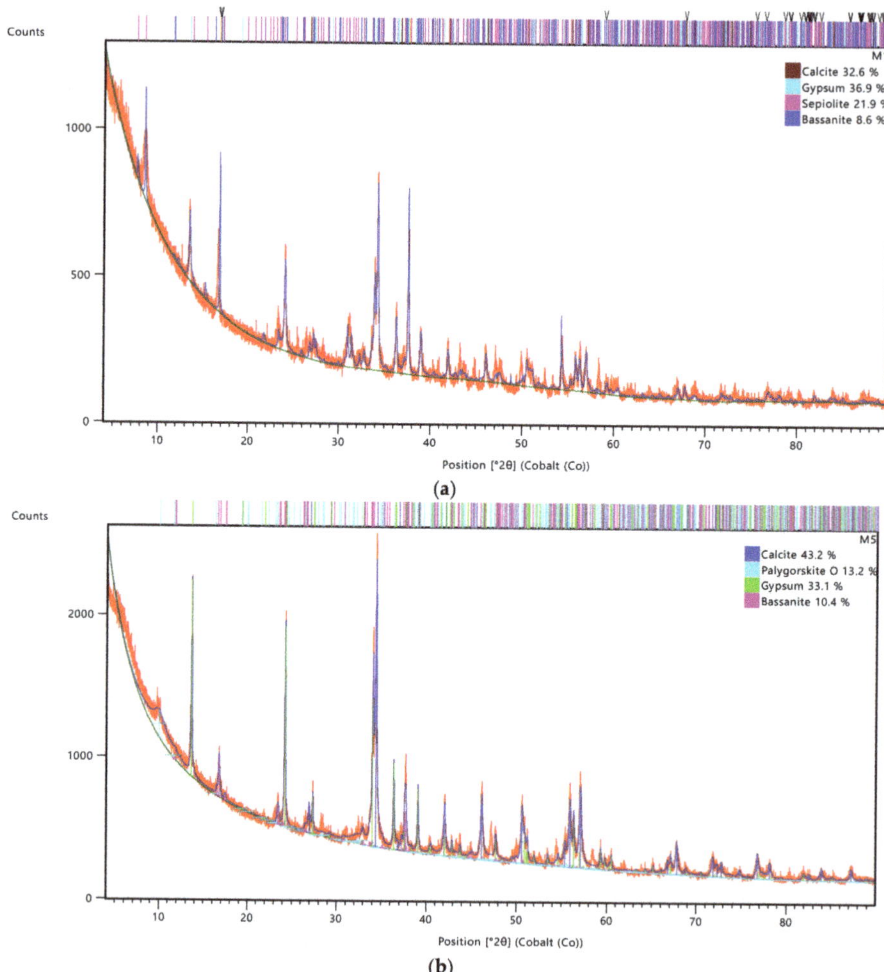

Figure 9. Quantitative results obtained by Rietveld refinement method; (**a**) sample M1 (GOF = 0.967); (**b**) sample M5 (GOF = 0.953); GOF = Goodness of fit.

The cell parameters obtained from the Rietveld refinement of M5 palygorskite are: **a** = 17.86, **b** = 5.21 Å, **c** = 12.77 Å, $\alpha = \beta = \gamma = 90°$. These data match very well with those of the diffraction pattern (ICSD 98-004-0688) used to identify this mineral (Table 6).

Table 6. Cell parameters for sepiolite (sample M1) and palygorskite (sample M5) calculated by Rietveld refinement method.

Sample	Mineral	a	b	c	α	β	γ	V(Å^3)
Blue/Ocuituco (M1)	Sepiolite	5.28(5)	13.40(6)	26.80 (1)	90°	90°	90°	1898
Dark blue/Oaxtepec (M5)	Palygorskite	17.86 (9)	5.21(6)	12.77(4)	90°	90°	90°	1189

The cell parameters obtained from the Rietveld refinement of M1 sepiolite are: **a** = 5.28 Å, **b** = 13.40 Å, **c** = 26.80 Å, $\alpha = \beta = \gamma = 90°$. These data match very well with those of the diffraction pattern (ICSD 98-003 1142) used to identify this mineral (Table 6).

4. Discussion

4.1. Chemical and Mineralogical Identification of Maya Blue

The majority of the samples analyzed with XRF fulfill the elemental criteria to be composed of Maya blue. Al of them have Si, Al and Mg.

In addition, when the elemental composition of XRF and the mineral composition of the blue samples are put together and compared (see Appendix A, where a table with direct comparison can be found), it can be observed that, even when the samples contained high values of Fe compared with those of the stucco, palygorskite was found. This is, as mentioned above, an almost certain indication that the blues were made with Maya blue, as shown in samples from Oaxtepec and Tezontepec, for example. Thus, high values of Fe acquired with non-destructive techniques should not be immediately attributed to the presence of Prussian blue. Furthermore, on the ubiquitous presence of Fe in these blue samples, several analyses of Maya blue suggest that Fe in the pigment replaces Mg and/or Al ions in the octahedral layer of the palygorskite, which occurs in all natural palygorskites [11]. However, as the pictorial layers and the stucco layers are so thin, XRF could be acquiring data from the plaster layer, and some quantity of the Fe could be due to the aggregates in the mixture with binders.

Nevertheless, other analyses had to be completed to confirm the presence of Maya blue conclusively. Using SEM-EDS, it was observed that all samples had the correct elemental composition and micromorphology to have a palygorskite-based pigment and thus the probability of these blue samples containing Maya blue.

Furthermore, samples studied by XRD exhibited palygorskite or sepiolite, which is a solid and unmistakable confirmation that Maya blue was used in the elaboration of the blue pigments employed in the wall paintings of these convents. Of the five pigment samples studied, four are characterized by the presence of palygorskite and one (M1) by the presence of sepiolite. In no sample have both minerals been found mixed (Figure 10).

Several palygorskites of great purity have been collected in several regions of the Yucatán Peninsula, all located in a radius of 40 km around the archaeological site of Uxmal [34]. The genesis of these deposits has been studied by several authors [35,36], and the most accepted interpretation in relation to their origin is that they were formed by direct crystallization in saline lagoons and on the shallow sea floor of the Yucatán Platform. The palygorskite–sepiolite clays of the Yucatan Peninsula are interbedded with limestone and dolomites.

The most reported mineral in the Yucatán Peninsula is palygorskite [36,37]; however, the presence of sepiolite has also been mentioned by multiple authors [35,38], but it has never been described in the deposits that emerge on the surface—those that were used by the Mayans in the manufacture of the Maya blue pigment. This is why sepiolite has never been identified in archaeological Maya blue to date and why the manufacture of this pigment has been exclusively associated with the mineral palygorskite [34,36,37,39–42]. However, it is very interesting to mention that sepiolite was found in archaeological samples from The Great Temple in Tenochtitlan, corresponding to the Aztec Empire [43,44].

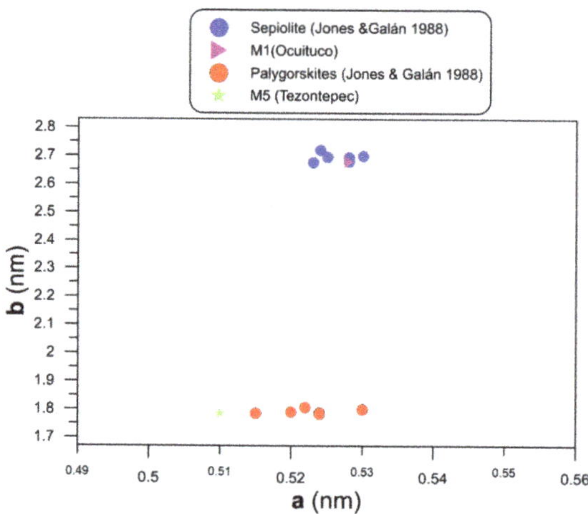

Figure 10. Cell parameters (**a,b**) for sepiolite of sample M1 (Ocuituco) and palygorskite of sample M5 (Tezontepec). Comparison with the cell parameters of the same minerals reported by Jones and Galán [45].

The mineralogical and elemental composition of the samples suggests that the palygorskite found in the blue samples of Oaxtepec and Tezontepec came from the Yucatan Peninsula and thus from the Mayan area; this evidence poses an interesting question about trading with a region that was not completely conquered in the XVI century. The sample from Ocuituco in which sepiolite was found has, with no doubt, another source that was not identified in this paper.

Finally, we want to emphasize the absence of pigments such as azurite or smalt. We know that their exclusion was not due to the lack of materials, since there was an active market for pigments from the European continent to New Spain and vice versa.

In the case of azurite from 1550 onwards, this blue pigment was a constant merchandise in the commercial trade between the Indies and Seville, and it was developed during the entire XVI century and a great part of the XVII century; it was extracted from mines in Santo Domingo and shipped in large quantities to Spain [20]. Moreover, azurite has been found widely in New Spain's oil canvases and wooden paintings [19]; its price varied between 34 and 54 maravedíes per pound [46].

On the other hand, in regard to blue smalt, Nicolás de Lambartengo, a neighbor of Seville, sent *"5 libras de esmalte de vidrio a 6 reales la libra, 1020 maravedíes"* in the ship Nuestra Señora de la Victoria in 1586 [46]. This report reveals small but constant quantities that were shipped, as well as their high cost, since at the time, a turkey cost approximately one real [47]; that is, to buy one pound, six turkeys would have been necessary. After azurite, blue smalt was the pigment most used by artists from New Spain, sometimes even mixed with azurite to achieve grayish shades [19].

In the case of mural painting, the quantities of pigments that were needed kept the Maya blue trade active. By maintaining the use of this pigment, not only were they using a specific material and tonality, but they kept the notions and meaning of this particular blue alive, becoming a suitable element for the integration of the indigenous culture to the western tradition.

4.2. Possible Interventions in the Wall Paintings

Samples from Cuernavaca, Tepoztlan, Tepetitlán, Acatlan, Epazoyucan, Oxtoticpac and Tepeji del Río probably had other pigments with/or instead of Maya blue, as shown by XRF analyses. The elemental composition showed elements such as cobalt or copper, which correspond to other blue pigments. These could be found for many reasons: (1) wall paintings were refreshed and re-painted later in the XVIII/XIX century; (2) wall paintings were possibly restored and have been subject to non-detectable (by eye) interventions; or (3) they have been restored in a way that can be clearly seen, as in Oxtoticpac.

In addition, when the structural formula of the clay minerals (palygorskite and sepiolite) was being calculated, an unusually high quantity of SiO_2 was noticed when compared with the quantity reported in the literature. For example, in the case of the blue sample of Tlaltizapan (Figure 11), it can be seen (both in spectra and in the image where a dark coating is present) that the surface is highly enriched with Si. The dark color points towards a lighter elemental composition than the rest of the material, reflecting the Si composition against the calcium carbonate and gypsum composition of the other layers.

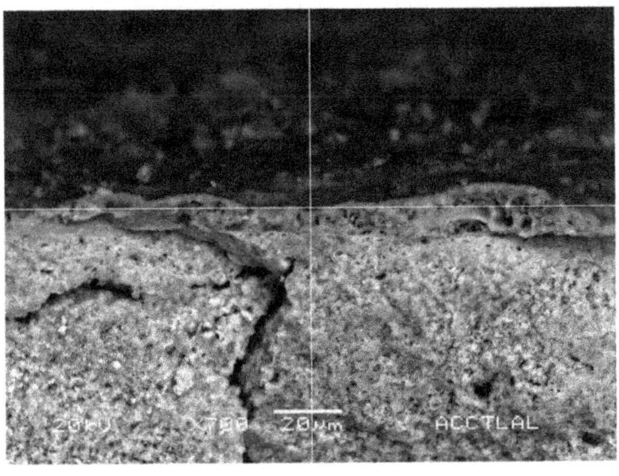
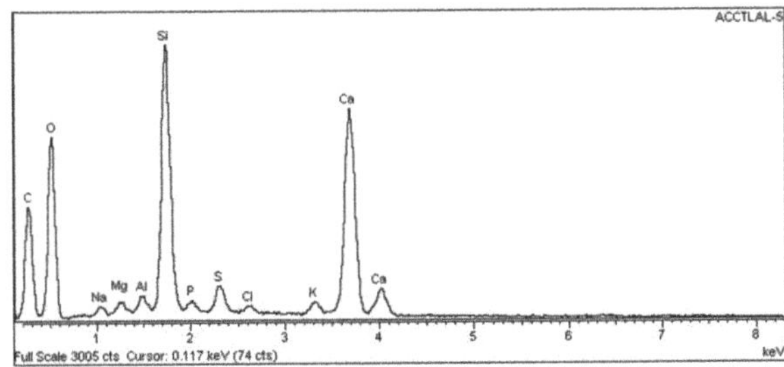

Figure 11. SEM image of Tlaltizapan blue. A dark coating enriched in Si can be observed at the surface.

We can explain this in two ways: firstly, that the wall paintings were submitted to a silicate-based consolidation process; and second, that in a given time, the wall paintings were subjected to a waterproofing treatment.

5. Conclusions

To conclude, Maya blue was apparently the preferred blue color for wall painting during the early and late XVI century, extending the use and life of a pre-Hispanic knowledge to the western Catholic tradition. It is noteworthy that none of the samples were painted al fresco, but with a variety of secco techniques. Moreover, almost every sample had a thin stucco or preparation base made of gypsum, a technique that was probably imported from Spain because the pre-Hispanic wall painting technique was almost always carried out with lime plasters and stuccos.

It is remarkable that the composition of palygorskite found in this research matches the compositions of those studied in the Yucatán Peninsula; this information indicates an active trade between different regions. In addition, it is important to highlight the presence of sepiolite in the blue sample from Ocuituco, as it corresponds to an entirely different tradition of pigments and perhaps even different trade routes, since there has not been any sepiolite found in Maya blues from the Mayan area.

Author Contributions: Conceptualization L.S.-M. and T.P.-P.; methodology, L.S.-M. and T.P.-P., A.F.-M., B.L.-T., K.C.-C., and M.F.U.-L.; investigation L.S.-M., T.P.-P., A.F.-M., B.L.-T., K.C.-C., and M.F.U.-L., resources, P.E.-G., A.F.-M., L.S.-M., and S.S.; data curation, L.S.-M. and T.P.-P.; writing—original draft preparation, L.S.-M., T.P.-P., A.F.-M., K.C.-C., and B.L.-T.; writing—review and editing, L.S.-M. and T.P.-P.; funding acquisition P.E.-G., A.F.-M., L.S.-M., and S.S. All authors have read and agreed to the published version of the manuscript.

Funding: This research was funded by PROGRAMA DE APOYO A PROYECTOS DE INVESTIGACIÓN E INNOVACIÓN TECNOLÓGICA (PAPIIT-UNAM) No. IN401918 and CONACYT: Laboratorios Nacionales, No. 299087.

Data Availability Statement: Data sharing not applicable.

Acknowledgments: The authors thank Mireia Lilit Solé Pi by the final English revision of the manuscript.

Conflicts of Interest: The authors declare no conflict of interest.

Appendix A

Convents with wall paintings are: Acatlán (Hidalgo), Acolman (Edo. Méx.), Actopan (Hidalgo), Alfajayucan (Hidalgo), Altihuetzia (Tlaxcala), Amecameca (Edo. Méx.), Apan (Hidalgo), Atlatlahucan (Morelos), Atlixco (Puebla), Atotonilco el Grande (Hidalgo), Ayotzingo (Edo. Méx.), Azcapotzalco (Cd. Mx.), Calpulalpan (Tlaxcala), Chimalhuacán-Chalco (Edo. Méx.), Cholula (San Gabriel) (Puebla), Churubusco (Cd. Mx.), Coyoacán (Cd. Mx.), Cuautinchán (Puebla), Cuernavaca (Morelos), Culhuacán (Cd. Mx.), Epazoyucan (Hidalgo), Huatlatlauca (Puebla), Huejotzingo (Puebla), Huexotla (Edo. Méx.), Ixmiquilpan (Hidalgo), Ixtapaluca (Edo. Méx.), Iztacalco (Cd. Mx.), Izúcar (Puebla), Juchitepec (Edo. Méx.), Malinalco(Edo. Méx.), Metepec (Edo. Méx.) Metztitlán (Hidalgo), Milpa Alta (Cd. Mx.), Molango (Hidalgo), Oaxtepec (Morelos), Ocuituco (Morelos), Oxtoticpac (Estado de México), San Jacinto Tenanitla (San Ángel) (Cd. Mx.), Singuilucan (Hidalgo), Tecali (Puebla), Tehuacán (Puebla), Tepeapulco (Hidalgo), Tepeji del Río (Hidalgo), Tepetitlán (Hidalgo), Tepetlaoxtoc (Edo. Méx.), Tepeyanco (Tlaxcala), Tepoztlán (Morelos), Tequixquiac (Hidalgo), Tetela del Volcán (Morelos), Tezontepec (Hidalgo), Tláhuac (Cd. Mx.), Tlahuelilpa (Hidalgo), Tlalmanalco (Edo. Méx.), Tlalnepantla (Edo. Méx.), Tlaltizapán (Morelos), Tlaquiltenango (Morelos), Tlaxcala (Tlaxcala), Tlayacapan (Mor.), Tochimilco (Puebla), Tula (Hidalgo), Xochimilco (Cd. Mx.), Yautepec (Morelos), Yecapixtla (Morelos), Zacatlán de las manzanas (Puebla), Zacualpan Amilpas (Morelos), Zempoala (Hidalgo) y Zinacantepec (Edo. Méx.) [48].

Appendix B

Elemental analysis of blue samples with XRF. The elements and percentage are shown in the columns and LE refers to light elements. Only the quantity of metals was intentionally tracked with XRF. The blue samples marked with * have lower quantities of Fe than stucco, samples marked with ~ had almost the same amount of Fe than stucco, the ones that are marked with ** have slightly more amount of Fe, and the blue samples marked with *** have higher amounts of Fe than the corresponding stucco.

Table A1. Elemental analysis obtained with XRF.

Convent	State	Hue	Ti	Mn	Fe	Co	Zr	Zn	Pb	V	Sn	Sb	Bi	Cr	Cu	Ni	As	LE
Yecapíxtla	Morelos	Blue ***	0.019	0.018	0.404	–	–	–	–	–	–	–	–	–	–	–	–	99.547
Yecapíxtla	Morelos	Stucco	0.029	–	0.189	–	–	–	–	–	–	–	–	–	–	–	–	99.761
Atlixco	Puebla	Blue ***	0.044	0.01	0.999	0.023	0.012	–	–	–	–	–	–	–	–	–	–	98.927
Atlixco	Puebla	Stucco	0.022	–	0.553	0.015	0.01	–	–	–	–	–	–	–	–	–	–	99.394
Ixmiquilpan	Hidalgo	Blue **	0.034	–	0.316	–	–	–	0.017	–	–	–	–	–	–	–	–	99.615
Ixmiquilpan	Hidalgo	Stucco	0.037	–	0.246	–	–	–	–	–	–	–	–	–	–	–	–	99.696
Cuernavaca	Morelos	light blue ***	0.026	–	0.515	–	0.015	0.102	0.018	–	–	–	–	–	–	–	–	99.306
Cuernavaca	Morelos	dark blue ***	0.079	–	0.341	0.011	0.014	–	0.012	0.028	–	–	–	–	–	–	–	99.498
Cuernavaca	Morelos	light blue ***	0.026	–	0.515	–	0.015	0.102	0.018	–	–	–	–	–	–	–	–	99.306
Cuernavaca	Morelos	Stucco	0.056	–	–	–	–	–	–	–	–	–	–	–	–	–	–	99.927
Oaxtepec	Morelos	blue/uper cloister~	0.024	–	0.077	–	–	–	–	–	–	–	–	–	–	–	–	99.881
Oaxtepec	Morelos	Stucco	0.012	–	0.079	–	–	–	–	–	–	–	–	–	–	–	–	99.893
Oaxtepec	Morelos	blue/lower cloister ***	0.04	0.011	0.439	–	–	–	–	–	–	–	–	–	–	–	–	99.498
Tepoztlan	Morelos	Blue ***	0.036	–	0.686	0.013	0.015	–	–	–	–	–	–	–	–	–	–	99.236
Tepoztlan	Morelos	Stucco	–	–	0.04	–	–	–	–	–	–	–	–	–	–	–	–	99.952
Ocuituco	Morelos	Blue **	0.017	–	0.078	–	–	–	–	–	–	–	–	–	–	–	–	99.897
Ocuituco	Morelos	Stucco	0.418	–	0.024	–	–	–	–	0.074	–	–	–	–	–	–	–	99.468
Zacualpan de Amilpas	Morelos	Blue~	0.046	0.016	0.876	0.024	0.015	–	–	–	–	–	–	–	–	–	–	99.016
Zacualpan de Amilpas	Morelos	Stucco	0.047	0.01	0.805	0.022	0.014	–	–	–	–	–	–	–	–	–	–	99.099
Tepeji del Río	Hidalgo	blue/background *	3.588	0.082	0.125	–	0.014	1.713	1.051	1.818	0.016	0.019	–	0.082	0.014	–	–	83.438
Tepeji del Río	Hidalgo	blue/virgin´s mantle *	2.375	0.043	0.163	–	–	1.8	4.47	1.112	0.027	0.021	–	0.044	0.037	–	0.347	89.55
Tepeji del Río	Hidalgo	Stucco	0.017	–	0.26	–	–	–	–	–	–	–	–	–	–	–	–	99.711
Alfajayucan	Hidalgo	Blue	0.012	–	0.067	–	–	–	0.108	–	–	–	–	–	–	–	0.036	99.766
Actopan	Hidalgo	blue/open chapel ***	0.03	–	0.892	–	–	–	–	–	–	–	–	–	–	–	–	99.056
Actopan	Hidalgo	Stucco	0.031	–	0.254	–	–	–	–	–	–	–	–	–	–	–	–	99.7
Tezontepec	Hidalgo	dark/blue **	0.023	–	0.308	–	–	–	–	–	–	–	–	–	–	–	–	99.645
Tezontepec	Hidalgo	Stucco	0.015	–	0.079	–	–	–	–	–	–	–	–	–	–	–	–	99.645
Tezontepec	Hidalgo	light/blue **	0.034	–	0.28	–	–	–	0.052	–	–	–	–	–	–	–	–	99.613
Tepetitlán	Hidalgo	blue/interior	0.084	0.01	0.708	0.015	0.02	–	–	–	–	–	–	–	–	–	–	99.155
Tepetitlán	Hidalgo	dark blue/superior part of wall painting	0.018	–	0.189	–	0.015	–	0.014	–	–	–	–	–	–	–	–	99.751
Tepetitlán	Hidalgo	light blue/superior part of wall painting	0.013	–	0.076	–	0.013	–	0.021	–	–	–	–	–	–	–	–	99.868
Tepetitlán	Hidalgo	blue/arch flowers	0.121	0.01	0.341	–	0.014	–	–	–	–	–	–	–	–	–	–	99.499
Actopan	Hidalgo	blue/cloister~	0.026	–	0.279	–	–	–	0.033	–	–	–	–	–	–	–	–	99.638
Actopan	Hidalgo	stucco/cloister	0.031	–	0.275	–	–	–	–	–	–	–	–	–	–	–	–	99.671
Acatlán	Hidalgo	Blue ***	0.069	0.074	0.567	0.01	0.011	–	–	–	–	–	–	–	–	–	–	99.266
Acatlán	Hidalgo	Stucco	0.033	0.05	0.28	–	–	–	–	–	–	–	–	–	–	–	–	99.28
Tlaltizapan	Morelos	blue/chapel ***	0.028	–	0.235	–	–	–	–	–	–	–	–	–	–	–	–	99.723
Tlaltizapan	Morelos	stucco/chapel	0.018	–	0.045	–	–	–	–	–	–	–	–	–	–	–	–	99.928
Tlaquiltenango	Morelos	Blue ***	–	–	0.331	–	–	–	–	–	–	–	–	–	–	–	–	99.659
Tlaquiltenango	Morelos	Stucco	–	–	0.068	–	–	–	–	–	–	–	–	–	–	–	–	99.922
Tepeapulco	Hidalgo	Blue ***	0.031	0.014	0.333	–	–	–	–	–	–	–	–	–	–	–	–	99.608
Tepeapulco	Hidalgo	Stucco	0.014	0.072	–	–	0.025	–	–	–	–	–	–	–	–	–	–	99.879
Zempoala	Hidalgo	Blue~	0.028	–	0.27	–	–	–	–	–	–	–	–	–	–	–	–	99.682
Zempoala	Hidalgo	Stucco	0.029	–	0.259	–	–	–	–	–	–	–	–	–	–	–	–	99.68
Epazoyucan	Hidalgo	light blue **	0.029	–	0.193	0.011	–	–	–	–	–	–	–	–	–	–	0.016	99.727
Epazoyucan	Hidalgo	dark blue **	0.018	–	0.256	–	–	–	0.092	–	–	–	–	–	–	–	0.018	99.593
Epazoyucan	Hidalgo	virgin mantle **	0.023	–	0.245	–	–	–	–	–	–	–	–	–	–	–	–	99.692
Epazoyucan	Hidalgo	Stucco	0.014	–	0.063	–	–	–	–	–	–	–	–	–	–	–	–	99.908
Oxtoticpac	Estado de México	Blue **	0.045	0.011	0.385	–	0.025	–	–	–	–	–	–	–	–	–	–	99.518
Oxtoticpac	Estado de México	blue/intervention **	0.043	–	0.393	0.249	0.024	–	0.012	–	–	–	0.028	–	–	0.023	0.099	99.108
Oxtoticpac	Estado de México	Stucco	0.021	–	0.203	–	0.021	–	–	–	–	–	–	–	–	–	–	99.7345
Cholula	Puebla	Greenish blue *	0.087	–	–	–	0.013	–	0.016	–	–	–	–	0.033	0.01	–	–	99.757
Cholula	Puebla	Greenish blue **	0.031	–	0.245	–	0.012	–	–	–	–	–	–	0.072	–	–	–	99.621
Cholula	Puebla	Stucco	0.013	–	0.054	–	0.013	–	–	–	–	–	–	–	–	–	–	99.918
Huejotzingo	Puebla	Blue	0.03	–	0.316	–	0.012	–	–	–	–	0.01	–	–	–	–	–	99.616
Huatlatlauca	Puebla	Blue~	0.037	–	0.283	–	–	–	–	–	–	–	–	–	–	–	–	99.646
Huatlatlauca	Puebla	Stucco	0.025	–	0.239	–	–	–	0.022	–	–	–	–	–	–	–	0.024	99.673
Tecali	Puebla	Blue **	0.033	–	0.42	–	0.031	–	–	–	–	–	–	–	–	–	–	99.485
Tecali	Puebla	Stucco	0.041	–	0.303	–	0.061	–	–	–	–	–	–	–	–	–	–	99.56
Cuauhtinchan	Puebla	dark blue/church **	0.019	0.02	0.448	–	0.011	–	–	–	–	–	–	–	–	–	–	99.492
Cuauhtinchan	Puebla	light blue/church **	0.066	0.014	0.423	–	0.012	–	–	–	–	–	–	–	–	–	–	99.479
Cuauhtinchan	Puebla	blue/convent *	0.023	–	0.189	–	–	–	–	–	–	–	–	–	–	–	–	99.773
Cuauhtinchan	Puebla	Stucco	0.024	0.017	0.284	–	0.011	–	–	–	–	–	–	–	–	–	–	99.658
Tula	Hidalgo	Blue **	0.126	–	0.303	–	–	0.025	–	–	–	–	–	–	–	–	–	99.513
Tula	Hidalgo	Stucco	0.022	–	0.25	–	0.016	–	–	–	–	–	–	–	–	–	–	99.626
Meztitlán	Hidalgo	Blue ***	0.039	–	0.465	–	0.016	0.256	0.02	–	–	–	–	0.398	–	–	–	98.793
Meztitlán	Hidalgo	stucco	0.015	–	0.062	–	–	–	–	–	–	–	–	–	–	–	–	99.911

Appendix A

Comparison between XRD and XRF analysis showing that even when the amount of Fe is high, the blue samples can be made by Maya blue. LE = Light elements.

Table A2. XRF data of blue samples cantaining palygorskite or sepiolite compared with XRF data of stucco without painting. Fe values can be compared.

Sample	Minerals Identified by XRD	Metallic Elements (XRF)	XRF (%)	Values of Fe
Blue Ocuituco (M1)	Calcite: $CaCO_3$ Gypsum: $CaSO_4\ 2H_2O$ Sepiolite:$Mg_4(Si_6O_{15})(OH)_2 \cdot 6H_2O$	Ti Fe LE	0.017 0.078 99.897	Fe values slightly higher
Stucco Ocuituco	–	Ti V Fe LE	0.418 0.074 0.024 99.468	
Light blue Tezontepec (M2)	Calcite: $CaCO_3$ Gypsum: $CaSO_4 \cdot 2H_2O$ Palygorskite: $(MgAl)_2Si_4O_{10}(OH) \cdot 4(H_2O)$	Ti Fe Pb LE	0.034 0.28 0.052 99.613	Higher values of Fe
Dark blue Tezontepec (M5)	Calcite: $CaCO_3$ Gypsum: $CaSO_4 \cdot 2H_2O$ Palygorskite:$(MgAl)_2Si_4O_{10}(OH) \cdot 4(H_2O)$	Ti Fe LE	0.023 0.308 99.645	
Stucco Tezontepec	–	Ti Fe LE	0.015 0.079 99.645	
Blue upper cloister Oaxtepec (M3)	Calcite: $CaCO_3$ Gypsum,: $CaSO_4 \cdot 2H_2O$ Palygorskite: $(MgAl)_2Si_4O_{10}(OH) \cdot 4(H_2O)$	Ti Fe LE	0.024 0.077 99.881	Lower values of Fe
Blue lower cloister Oaxtepec (M4)	Calcite: $CaCO_3$ Palygorskite: $(MgAl)_2Si_4O_{10}(OH) \cdot 4(H_2O)$	Ti Mn Fe LE	0.04 0.011 0.439 99.498	Higher values of Fe
Stucco Oaxtepec	–	Ti Fe LE	0.012 0.079 99.893	–

References

1. Flores Moran, A. El Sincretismo Cultural y La Conquista de La Imagen: Las Normatividades Hispana y Náhuatl Vistas En La Pintura Mural Del Siglo XVI. Bachelor's Thesis, Escuela Nacional de Antropología e Historia, Mexico City, Mexico, 2011.
2. Reyes-Valerio, C. La pintura mural del siglo xvi en Mexico. *Jahrb. Gesch. Lat.* **1983**, *20*. [CrossRef]
3. Reyes Valerio, C. *El Pintor de Conventos. Los Murales Del Siglo XVI En La Nueva Españao Title*; Instituto Nacional de Antropología e Historia: Mexico City, Mexico, 1989.
4. Escalante Gonzalbo, P. El Patrocinio Del Arte Indocristiano. In *Patrocinio, Colección y Circulación de las Artes*; Curiel Méndez, G.A., Ed.; Instituto de Investigaciones Estéticas, UNAM: Mexico City, Mexico, 1997.
5. Kubler, G. *Arquitectura Mexicana Del Siglo XVI*; Fondo de Cultura Económica: Mexico City, Mexico, 2012.
6. Estrada Gerlero, I. La Pintura Mural Durante El Virreinato. In *En Historia del Arte Mexicano: Arte Colonial*; Manrique, J., Vargaslugo, E., Eds.; Secretaría de Educación pública, Instituto nacional de Bellas Artes: Mexico City, Mexico, 1982; Volume 6, pp. 1–9.
7. Post, J.E.; Heaney, P.J. Synchrotron powder X-ray diffraction study of the structure and dehydration behavior of palygorskite. *Am. Mineral.* **2008**, *93*, 667–675. [CrossRef]
8. Sánchez del Río, M.; Martinetto, P.; Solís, C.; Reyes-Valerio, C. PIXE Analysis on Maya blue in prehispanic and colonial mural paintings. *Nucl. Instrum. Methods Phys. Res. Sect. B* **2006**, *249*, 628–632. [CrossRef]
9. Sánchez del Río, M.; Doménech, A.; Doménech-Carbó, M.T.; De Agredos Pascual, M.L.V.; Suárez, M.; García-Romero, E. The Maya Blue Pigment. In *Developments in Clay Science*; Elsevier: Amsterdam, The Netherlands, 2011; Volume 3, pp. 453–481. [CrossRef]
10. Tagle, A.A.; Paschinger, H.; Richard, H.; Infante, G. Maya Blue: Its presence in cuban colonial wall paintings. *Stud. Conserv.* **1990**, *35*, 156–159. [CrossRef]
11. Sánchez del Río, M.; Reyes Valerio, C.; Picquart, M.; Haro-Poniatowski, E.; Lima, E.; Lara, V.H.; Castillo, P.; Vázquez, H.; Uc, V.H.; Páez, S.; et al. Nuevas Investigaciones Sobre El Azul Maya. In *La Ciencia de Materiales y su Impacto en la Arqueología*; Mendoza, D., Arenas, J.A., Ruvalcaba, J.L., Rodríguez, V., Eds.; Innovación Editorial Lagares de México: Puebla, Mexico, 2006; Volume III, pp. 11–20.

12. Sanz, E.; Arteaga, A.; García, M.A.; Cámara, C.; Dietz, C. Chromatographic analysis of indigo from Maya blue by LC-DAD-QTOF. *J. Archaeol. Sci.* **2012**, *39*, 3516–3523. [CrossRef]

13. Haude, M.E. Identification of colorants on maps from the early colonial period of new Spain (Mexico). *J. Am. Inst. Conserv.* **1998**, *37*, 240–270. [CrossRef]

14. García-Bucio, M.A. Espectroscopía Raman Para Estudio No Destructivo de Pigmentos y Colorantes. Master´s Thesis, Universidad Nacional Autónoma de México, Mexico City, Mexico, 2015.

15. Sánchez Del Río, M.; Martinetto, P.; Somogyi, A.; Reyes-Valerio, C.; Dooryhée, E.; Peltier, N.; Alianelli, L.; Moignard, B.; Pichon, L.; Calligaro, T.; et al. Microanalysis Study of Archaeological Mural Samples Containing Maya Blue Pigment. *Spectrochim. Acta Part B At. Spectrosc.* **2004**, *59*, 1619–1625. [CrossRef]

16. Casanova González, E. Espectroscopía Raman y SERS En El Estudio Del Patrimonio Cultural Mexicano. Ph.D. Thesis, Universidad Nacional Autónoma de México, Mexico City, Mexico, 2012.

17. Wong Rueda, M. Análisis No Destructivos Para Caracterización In Situ de Pintura Mural Colonial. Bahelor's Thesis, Universidad Nacional Autónoma de México, Mexico City, Mexico, 2013.

18. Doerner, M. *Los Materiales de Pintura y Su Empleo En El Arte*; Reverte: Barcelona, Spain, 1986.

19. Arroyo, E.; Espinosa, M.E.; Falcón, T.; Hernández, E. Variaciones celestes para pintar el manto de la virgen. *An. Inst. Investig. Estéticas* **2012**, *34*, 85. [CrossRef]

20. Galán, R.B. La obtención de pigmentos azules para las obras de Felipe II: Comercio Europeo y Americano. In *Art Technology: Sources and Methods. Proceedings of the Second Symposium of the Art Technological Source Research Working Group*; Kroustallis, S., Bellestrem, A., Eds.; Museo Nacional Centro de Arte Reina Sofía: Madrid, Spain, 2008; pp. 55–63. ISBN 10:1904982298.

21. Gil, M.; Serrão, V.; Carvalho, M.L.; Longelin, S.; Dias, L.; Cardoso, A.; Caldeira, A.T.; Rosado, T.; Mirão, J.; Candeias, A.E. Material and diagnostic characterization of 17th century mural paintings by spectra-colorimetry and SEM-EDS: An insight look at josé de escovar workshop at the CONVENT of Na Sra Da Saudação (Southern Portugal). *Color Res. Appl.* **2014**, *39*, 288–306. [CrossRef]

22. Piovesan, R.; Mazzoli, C.; Maritan, L.; Cornale, P. Fresco and Lime-Paint: An experimental study and objective criteria for distinguishing between these painting techniques. *Archaeometry* **2012**, *54*, 723–736. [CrossRef]

23. Guzmán, F.; Maier, M.; Pereira, M.; Sepúlveda, M.; Siracusano, G.; Cárcamo, J.; Castellanos, D.; Gutiérrez, S.; Tomasini, E.; Corti, P.; et al. Programa iconográfico y material en las pinturas murales de la iglesia de san andrés de pachama, Chile. *Colon. Latin Am. Rev.* **2016**, *25*, 245–264. [CrossRef]

24. Marey Mahmoud, H.; Kantiranis, N.; Ali, M.; Stratis, J. Characterization of ancient egyptian wall paintings, the excavations of Cairo University at Saqqara. *Int. J. Conserv. Sci.* **2011**, *2*, 145–154.

25. Laclavetine, K.; Ruvalcaba-Sil, J.L.; Duverger, C.; Melgar, E. Arqueometría: Aplicación de La Física En Arqueología. Aportes de La Fluorescencia de Rayos X (XRF) En El Estudio de Artefactos Aztecas. In *2nda Reunión Científica sobre Innovación y Ciencia Aplicada al Estudio, Conservación y Restauración del Patrimonio Cultural*; University of the Basque Country: Bilbao, Spain, 2013; pp. 1–5. [CrossRef]

26. Spores, J. Analisis de Procedencia y Estudio de Artefactos de Obsidiana Del Proyecto Arqueológico Pueblo Viejo de Teposcolula Yucundaa. In *Yucundaa. La Ciudad Mixteca Yucundaa-Pueblo Viejo de Teposcolula, Oaxaca y Su Transformacion Prehispanica-Colonial*; Spores, R., Robles García, N.M., Eds.; Instituto Nacional de Antropología e Historia, Fundación Alfredo Harp Helú: Mexico City, Mexico, 2014; pp. 541–560.

27. Liritzis, I.; Zacharias, N. Portable XRF of Archaeological Artifacts: Current Research, Potentials and Limitations. In *X-ray Fluorescence Spectrometry (XRF) in Geoarchaeology*; Shackley, M.S., Ed.; Springer: New York, NY, USA, 2011; pp. 109–142. [CrossRef]

28. Desnica, V.; Škarić, K.; Jembrih-Simbuerger, D.; Fazinić, S.; Jakšić, M.; Mudronja, D.; Pavličić, M.; Peranić, I.; Schreiner, M. Portable XRF as a valuable device for preliminary in situ pigment investigation of wooden inventory in the trski vrh church in Croatia. *Appl. Phys. A Mater. Sci. Process.* **2008**, *92*, 19–23. [CrossRef]

29. Robador, M.D.; De Viguerie, L.; Pérez-Rodríguez, J.L.; Rousselière, H.; Walter, P.; Castaing, J. The structure and chemical composition of wall paintings from islamic and christian times in the seville alcazar. *Archaeometry* **2016**, *58*, 255–270. [CrossRef]

30. Appoloni, C.R.; Blonski, M.S.; Parreira, P.S.; Souza, L.A.C. Pigments elementary chemical composition study of a gainsborough attributed painting employing a portable X-rays fluorescence system. *AIP Conf. Proc.* **2007**, *884*, 459–464. [CrossRef]

31. Rietveld, H. A profile refinement method for nuclear and magnetic structures. *J. Appl. Crystallogr.* **1969**, *2*, 65–71. [CrossRef]

32. Moore, D.M.; Reynolds, R.C. *X-ray Diffraction and the Identification and Analysis of Clay Minerals*, 2nd ed.; Oxford University Press: Oxford, UK; New York, NY, USA, 1997.

33. García-Romero, E.; Suarez, M. On the chemical composition of sepiolite and palygorskite. *Clays Clay Miner.* **2010**, *58*, 1–20. [CrossRef]

34. Sánchez del Río, M.; Suarez, M.; García-Romero, E. The occurrence of palygorskite in the yucatán peninsula: Ethno-historic and archaeological contexts. *Archaeometry* **2009**, *51*, 214–230. [CrossRef]

35. Isphording, W.C.; Wilson, E.M. The relationship of "volcanic ash," sakulum, and palygorskite northern yucatan Maya ceramics. *Am. Antiq.* **1974**, *39*, 483–488. [CrossRef]

36. De Pablo Galán, L. Palygorskite in eocene-oligocene lagoonal environment, Yucatán, Mexico. *Rev. Mex. Cienc. Geológicas* **1996**, *13*, 94–104.

37. Arnold, D.E. Ethnomineralogy of Ticul, Yucatan Potters: Etics and emics. *Am. Antiq.* **1971**, *36*, 20–40. [CrossRef]

38. Stinnesbeck, W.; Keller, G.; Adatte, T.; Harting, M.; Stüben, D.; Istrate, G.; Kramar, U. Yaxcopoil-1 and the chicxulub impact. *Int. J. Earth Sci.* **2003**, *93*. [CrossRef]

39. Arnold, D.; Bohor, B. Attapulgite and Maya blue: An ancient mine comes to light. *Archaeology* **1975**, *28*, 22–29.
40. Folan, W.J. Sacalum, Yucatán: A Pre-hispanic and contemporary source of attapulgite. *Am. Antiq.* **1969**, *34*, 182–183. [CrossRef]
41. Arnold, D.E. Maya blue and palygorskite: A second possible pre-columbian source. *Anc. Mesoam.* **2005**, *16*, 51–62. [CrossRef]
42. Arnold, D.E.; Neff, H.; Glascock, M.D.; Speakman, R.J. Sourcing the palygorskite used in Maya blue: A pilot study comparing the results of INAA and LA-ICP-MS. *Lat. Am. Antiq.* **2007**, *18*, 44–58. [CrossRef]
43. Ortega Avilés, M. Caracterización de Pigmentos Prehispánicos Por Técnicas Analíticas Modernas. Ph.D. Thesis, Universidad Autónoma del Estado de México, Toluca, Mexico, 2003.
44. Shepard, A.O.; Gottlieb, H.B. *Maya Blue: Alternative Hypothesis. Notes from a Ceramic Laboratory*; Carnegie Institution of Washington: Washington, DC, USA, 1962.
45. Jones, B.F.; Galán, E. Sepiolite and Palygorskite. In *Reviews in Mineralogy. Hydrous Phyllosilicates*; Bailey, S.W., Ed.; Mineralogical Society of America: Madison, WI, USA, 1988; Volume 19, pp. 631–674.
46. Sánchez, J.M.; Quiñones, M.D. Materiales Pictóricos Enviados a América En El Siglo XVI. *An. Inst. Investig. Estéticas XXXI* **2009**, *95*, 45–67. [CrossRef]
47. Reyes García, L. Cómo Te Confundes? Acaso No Somos Conquistados?: Anales de Juan Bautista. In *Anales de Juan Bautista*; Centro de Investigaciones y Estudios Superiores en Antropología Social, Biblioteca Lorenzo Boturini, Insigne y Nacional Basílica de Guadalupe: Mexico City, Mexico, 2001; 343p.
48. Flores Morán, A. Cambios y Continuidades de La Pintura Mural Conventual Del Altiplano Central (1521–1640). Orígenes, Tradiciones, Técnicas y Estilos. Ph.D. Thesis, Universidad Nacional Autónoma de México, Mexico City, Mexico, 2020.

MDPI

St. Alban-Anlage 66

4052 Basel

Switzerland

Tel. +41 61 683 77 34

Fax +41 61 302 89 18

www.mdpi.com

Coatings Editorial Office

E-mail: coatings@mdpi.com

www.mdpi.com/journal/coatings

www.ingramcontent.com/pod-product-compliance
Lightning Source LLC
LaVergne TN
LVHW070848170726
843515LV00005B/599